Everyday Math DeMYSTiFieD®

DeMYSTiFieD® Series

Accounting Demystified
Advanced Calculus Demystified
Advanced Physics Demystified
Advanced Statistics Demystified
Algebra Demystified
Alternative Energy Demystified
Anatomy Demystified
Astronomy Demystified
Audio Demystified
Biology Demystified
Biotechnology Demystified
Business Calculus Demystified
Business Math Demystified
Business Statistics Demystified
C++ Demystified
Calculus Demystified
Chemistry Demystified
Chinese Demystified
Circuit Analysis Demystified
College Algebra Demystified
Corporate Finance Demystified
Databases Demystified
Data Structures Demystified
Differential Equations Demystified
Digital Electronics Demystified
Earth Science Demystified
Electricity Demystified
Electronics Demystified
Engineering Statistics Demystified
Environmental Science Demystified
Ethics Demystified
Everyday Math Demystified
Fertility Demystified
Financial Planning Demystified
Forensics Demystified
French Demystified
Genetics Demystified
Geometry Demystified
German Demystified
Home Networking Demystified
Investing Demystified
Italian Demystified
Japanese Demystified
Java Demystified
JavaScript Demystified
Lean Six Sigma Demystified
Linear Algebra Demystified
Logic Demystified
Macroeconomics Demystified
Management Accounting Demystified
Math Proofs Demystified
Math Word Problems Demystified
MATLAB® Demystified
Medical Billing and Coding Demystified
Medical Terminology Demystified
Meteorology Demystified
Microbiology Demystified
Microeconomics Demystified
Nanotechnology Demystified
Nurse Management Demystified
OOP Demystified
Options Demystified
Organic Chemistry Demystified
Personal Computing Demystified
Pharmacology Demystified
Philosophy Demystified
Physics Demystified
Physiology Demystified
Pre-Algebra Demystified
Precalculus Demystified
Probability Demystified
Project Management Demystified
Psychology Demystified
Quality Management Demystified
Quantum Mechanics Demystified
Real Estate Math Demystified
Relativity Demystified
Robotics Demystified
Sales Management Demystified
Signals and Systems Demystified
Six Sigma Demystified
Spanish Demystified
sql Demystified
Statics and Dynamics Demystified
Statistics Demystified
Technical Analysis Demystified
Technical Math Demystified
Trigonometry Demystified

Everyday Math

DeMYSTiFieD®

Stan Gibilisco

Second Edition

New York Chicago San Francisco Lisbon London Madrid Mexico City
Milan New Delhi San Juan Seoul Singapore Sydney Toronto

McGraw-Hill books are available at special quantity discounts to use as premiums and sales promotions, or for use in corporate training programs. To contact a representative, please e-mail us at bulksales@mcgraw-hill.com.

Everyday Math DeMYSTiFieD®, Second Edition

1 2 3 4 5 6 7 8 9 0 DOC/DOC 1 2 0 9 8 7 6 5 4 3 2

ISBN 978-0-07-179013-0
MHID 0-07-179013-6

Sponsoring Editor
Judy Bass

Acquisitions Coordinator
Bridget L. Thoreson

Editing Supervisor
David E. Fogarty

Production Supervisor
Pamela A. Pelton

Project Managers
Nancy Dimitry,
Joanna Pomeranz,
D&P Editorial Services

Composition
D&P Editorial Services

Copy Editor
Nancy Dimitry,
D&P Editorial Services

Proofreader
Don Dimitry,
D&P Editorial Services

Art Director, Cover
Jeff Weeks

Cover Illustration
Lance Lekander

To Tony, Samuel, and Tim

About the Author

Stan Gibilisco, an electronics engineer and mathematician, has authored multiple titles for the McGraw-Hill *Demystified* and *Know-It-All* series, along with numerous other technical books and dozens of magazine articles. His work has been published in several languages.

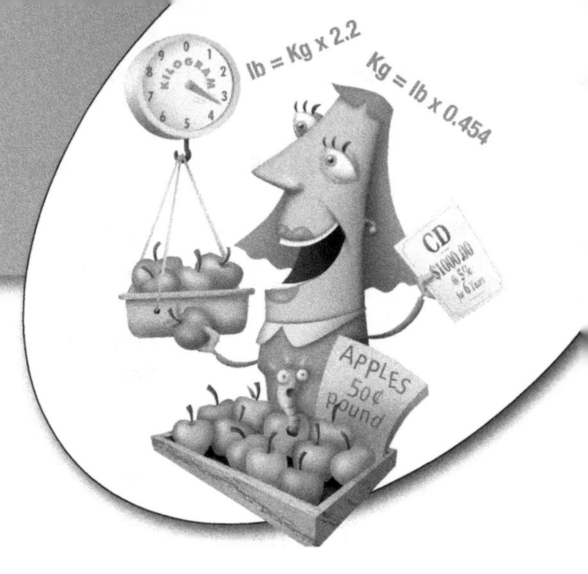

Contents

Introduction

This book can help you learn or review the fundamentals of "real-world" mathematics without taking a formal course. It can also serve as a supplemental text in a classroom, tutored, or home-schooling environment. Nothing in this book goes beyond the high-school level. If you want to explore any of these subjects in more detail, you can select from several *Demystified* books dedicated to mathematics topics.

How to Use This Book

As you take this course, you'll find an "open-book" multiple-choice quiz at the end of every chapter. You may (and should) refer to the chapter text when taking these quizzes. Write down your answers, and then give your list of answers to a friend. Have your friend tell you your score, but not which questions you missed. The correct answer choices are listed in the back of the book. Stay with a chapter until you get most of the quiz answers correct.

The course concludes with a final exam. Take it after you've finished all the chapters and taken their quizzes. You'll find the correct answer choices listed in the back of the book. With the final exam, as with the quizzes, have a friend reveal your score without letting you know which questions you missed. That way, you won't subconsciously memorize the answers. You might want to take the final exam two or three times. When you get a score that makes you happy, you can (and should) check to see where your strengths and weaknesses lie.

I've posted explanations for the chapter-quiz answers (but not for the final-exam answers) on the Internet. As we all know, Internet particulars change; but if you conduct a phrase search on "Stan Gibilisco," you should get my Web site as one of the first hits. You'll find a link to the explanations there. As of this writing, the site location is

www.sciencewriter.net

Strive to complete one chapter every two or three weeks. Don't rush, but don't go too slowly either. Proceed at a steady pace and keep it up. That way, you'll complete the course in a few months. (As much as we all wish otherwise, nothing can substitute for "good study habits.") After you finish this book, you can use it as a permanent reference.

I welcome your ideas and suggestions for future editions.

Stan Gibilisco

Everyday Math

DeMYSTiFieD®

chapter 1

Numerals

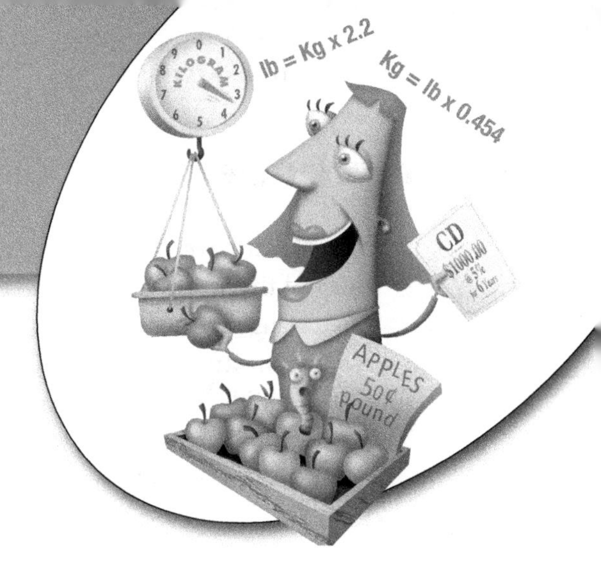

A *number* reveals a quantity, an amount, or an extent, but it's not a material thing. You can imagine a number, but you can't see one. In contrast, a *numeral* is a visible symbol (or group of symbols) that represents a number. Since the beginning of civilization, people have invented various numeral systems.

CHAPTER OBJECTIVES

In this chapter, you will

- Portray numbers in pictorial form.
- See how the ancient Romans expressed quantities as groups of letters.
- Learn how the Arabic numeration system works.
- Discover alternative counting methods.
- Convert quantities between different numeration systems.

Numeric Pictures

When you have a lot of things to count, you can arrange them into groups. In most of the world, people use the quantity we call *ten* as the basis for counting. Historians might debate how this idea got so well established. Maybe it arose when primitive people used the appendages on their hands to tally things up, just as children (and sometimes even I) still do. Based on that notion, you can portray the numeral representing ten items by writing down F's for "fingers" and T's for "thumbs" as

FFFFT (left hand)
TFFFF (right hand)

If you refer to your fingers and thumbs collectively as "digits" and do away with the distinction between your left hand and your right hand, you can denote the same quantity using D's for "digits" as

DDDDD
DDDDD

Mathematicians call the numbering scheme based on multiples of ten the *decimal numeration system*. The number ten is written as the numeral 10. Two groups of ten make *twenty* (20); three groups of ten make *thirty* (30); four groups of ten make *forty* (40).

Tens aren't the only *number base* (also called *radix*) that people employ when they want to describe things in terms of their quantity. People occasionally count things in *dozens* or groups of *twelve*. Your friendly local mathematician would call this scheme the *duodecimal numeration system*. You can show the concept of twelve items unambiguously using I's (for "items") as

IIIIII
IIIIII

When you have more than ten items and you want to represent the quantity in the decimal system, you state the number of complete tens with the extras left over. For example:

- The numeral 26 means two groups of ten along with six more
- The numeral 30 means three groups of ten along with no more
- The numeral 89 means eight groups of ten along with nine more
- The numeral 90 means nine groups of ten along with no more
- The numeral 99 means nine groups of ten along with nine more

- The numeral 100 means ten groups of ten along with no more
- The numeral 101 means ten groups of ten along with one more

TIP *You can render numerals for increasingly large numbers by thinking about how you would pack things into trays or boxes. A tray might hold 12 rows of 12 apples, for example. Then you'd have room for 12 × 12, or 144 apples. That's an example of how you can expand the duodecimal system, getting a dozen-dozen, also known as a* gross*. You might stack 12 trays of 144 apples, one on top of the other, in a cube-shaped box holding 12 × 12 × 12 apples. If you go to all that trouble, you'll end up with 1728 apples. (I don't know of any specific term for that quantity. Maybe we can call it a "hypergross" or a "grozen.")*

PROBLEM 1-1

Draw a picture showing how you can divide a single large square into 100 smaller squares.

SOLUTION

You can break the large square into 10 rows and 10 columns, as shown in Fig. 1-1. Each row contains 10 squares going across from left to right. Each column contains 10 squares going down from top to bottom. The total equals 10 × 10, or 100 small squares.

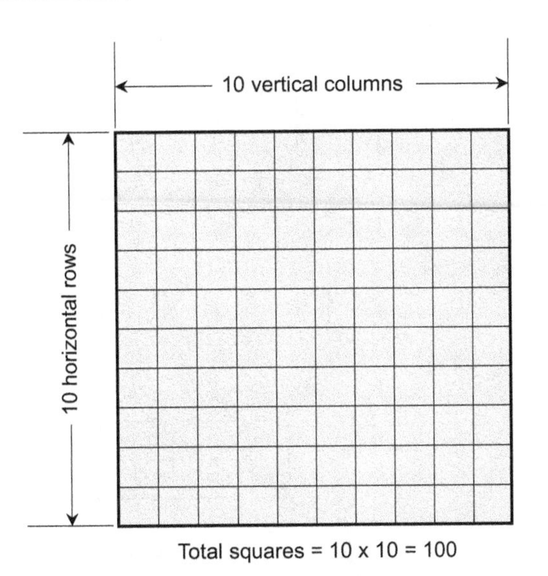

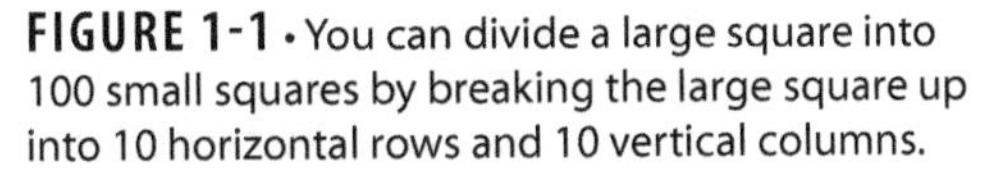
FIGURE 1-1 • You can divide a large square into 100 small squares by breaking the large square up into 10 horizontal rows and 10 vertical columns.

PROBLEM 1-2

Draw two different pictures showing how you can portray the numeral 1000 as tiny squares within larger squares, all inside a massive rectangle.

SOLUTION

You can create 10 identical "copies" of the large square from Fig. 1-1, and then assemble them in a single horizontal row or a single vertical column to create a huge rectangle such as the one shown in Fig. 1-2A. Then you'll have $10 \times 10 \times 10 = 1000$ tiny squares. Alternatively, you can arrange the 10 duplicates of Fig. 1-1 into two rows of five large squares each, or two columns of five large squares each, to obtain a huge rectangle such as the one in Fig. 1-2B. In either case, you have a total of $2 \times 5 \times 10 \times 10 = 1000$ tiny squares.

PROBLEM 1-3

Draw a picture that illustrates the concept of 10,000 as tiny squares within larger squares, all inside a massive square.

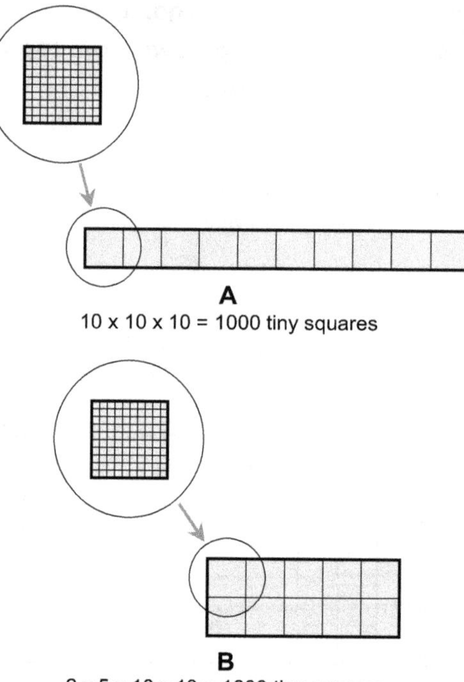

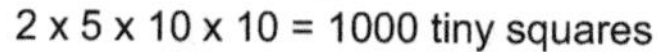

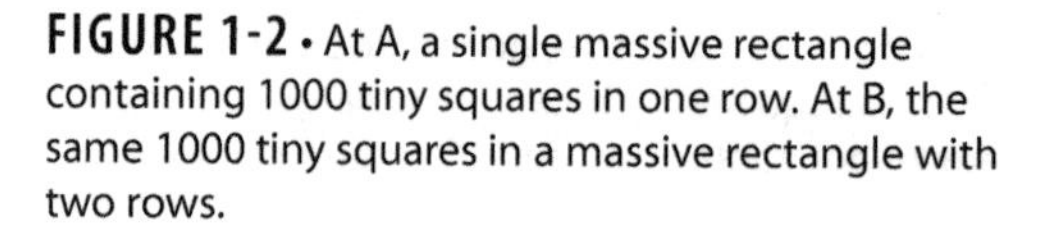

FIGURE 1-2 • At A, a single massive rectangle containing 1000 tiny squares in one row. At B, the same 1000 tiny squares in a massive rectangle with two rows.

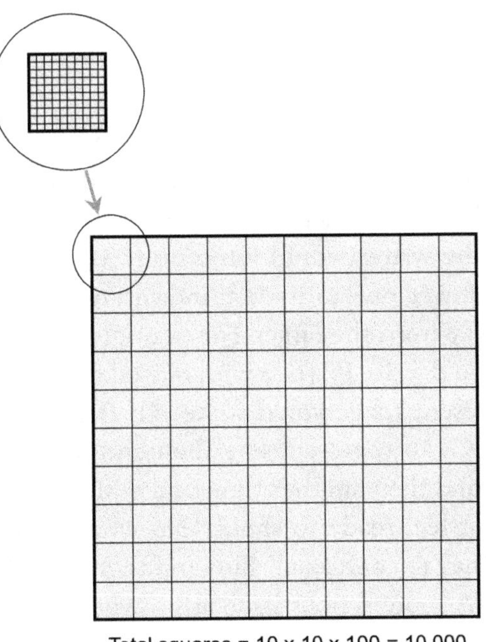

FIGURE 1-3 • You can assemble 100 large squares, each containing 100 tiny squares, into a massive array of 10,000 tiny squares.

SOLUTION

You can create 100 identical "copies" of the large square from Fig. 1-1, and then assemble them just as you did with the original tiny squares, getting 10 rows of 10 large squares each. Figure 1-3 shows the result.

Roman Numerals

The counting scheme described at the start of this chapter resembles a system that much of the world worked with until a few centuries ago: the *Roman numeration system*, more often called *Roman numerals*. We use uppercase letters of the alphabet to represent quantities as follows:

- I means one
- V means five
- X means ten
- L means fifty
- C means a hundred

- D means five hundred
- M usually represents a thousand
- K sometimes represents a thousand

The people who designed the Roman system didn't like to put down more than three identical symbols in a row. Instead of putting four identical symbols one after another, the writer would jump up to the next higher symbol and then put the next lower one to its left, indicating that the smaller quantity should be taken away from the larger. For example, instead of IIII (four ones) to represent four, you'd write IV (five with one taken away). Instead of XXXX (four tens) to represent forty, you'd write XL (fifty with ten taken away). Instead of MDXXXX to represent one thousand nine hundred, you'd write MCM (a thousand and then another thousand with a hundred taken away).

By now you must feel ready to shout, "No wonder people got away from Roman numerals. They're confusing!" But confusion isn't the only trouble with the Roman system. A more serious issue arises when you try to do arithmetic. The Roman scheme gives you no way to express the quantity zero. This detail might not seem important at first thought. But when you start adding and subtracting, and especially when you start multiplying and dividing, you'll have a hard time getting along without zero. The numeral 0 serves as a *placeholder*, keeping the structure intact when you want to write any but the smallest numbers.

Still Struggling

Let's write down all the *counting numbers* from one to twenty-one as Roman numerals. This exercise will give you a sense of how the symbolic arrangements can represent adding-on or taking-away of quantities. The first three are easy: I means one, II means two, and III means three. To denote four, we write IV, meaning that we take one away from five. Proceeding along further, V means five, VI means six, VII means seven, and VIII means eight. To represent nine, we write IX, meaning that we take one from ten. Then X means ten, XI means eleven, XII means twelve, and XIII means thirteen. To denote fourteen, we write XIV, which means ten with four added on. Then XV means fifteen, XVI means sixteen, XVII means seventeen, and XVIII means eighteen. For nineteen, we write XIX, which means ten with nine more added on. Continuing, we have XX that stands for twenty, and XXI to represent twenty-one.

TABLE 1-1 Some examples of Roman numerals. From this progression, you can see how the system works for fairly large numbers.

1 = I	10 = X	100 = C	910 = CMX	991 = CMXCI
2 = II	20 = XX	200 = CC	920 = CMXX	992 = CMXCII
3 = III	30 = XXX	300 = CCC	930 = CMXXX	993 = CMXCIII
4 = IV	40 = XL	400 = CD	940 = CMXL	994 = CMXCIV
5 = V	50 = L	500 = D	950 = CML	995 = CMXCV
6 = VI	60 = LX	600 = DC	960 = CMLX	996 = CMXCVI
7 = VII	70 = LXX	700 = DCC	970 = CMLXX	997 = CMXCVII
8 = VIII	80 = LXXX	800 = DCCC	980 = CMLXXX	998 = CMXCVIII
9 = IX	90 = XC	900 = CM	990 = CMXC	999 = CMXCIX

PROBLEM 1-4

Write down some Roman numerals in a table. In the first column, put down the equivalents of one to nine in steps of one. In a second column, put down the equivalents of ten to ninety in steps of ten. In a third column, put down the equivalents of one hundred to nine hundred in steps of one hundred. In a fourth column, put down the equivalents of nine hundred ten to nine hundred ninety in steps of ten. In a fifth column, put down the equivalents of nine hundred ninety-one to nine hundred ninety-nine in steps of one.

SOLUTION

Refer to Table 1-1. The first column lies farthest to the left, and the fifth column lies farthest to the right. For increasing values in each column, read downward. "Normal" numerals appear along with their Roman equivalents.

Arabic Numerals

Mathematicians in southern Asia invented the numeration system that most of the world employs nowadays. During the two or three hundred years after the system's first implementation, invaders from the Middle East picked it up. (Good ideas have a way of catching on, even among invaders.) Eventually, most of the civilized world adopted the *Hindu-Arabic numeration system*. The "Hindu" part of the name comes from India, and the "Arabic" part comes from the Middle East. You'll often hear the symbols in this system referred to simply as *Arabic numerals*.

In an Arabic numeration system, every digit represents a quantity ranging from zero to nine: the familiar symbols 0, 1, 2, 3, 4, 5, 6, 7, 8, and 9. The original Hindu inventors of the system came up with an interesting way of expressing numbers larger than nine. They gave each digit more or less "weight" or value, depending on where it appeared in relation to other digits in the same numeral. These innovators got the idea that every digit in a numeral should have ten times the value of the digit (if any) to its right. When building up the numeric representation for a large number, there might be no need for a digit in a particular place, but a crucial need for a digit on either side of it—giving birth to the notion of 0, a numeral to represent nothing!

Figure 1-4 shows an example of a numeral that represents a large number. Note that the digit 0, also called a *cipher*, is just as important as any other digit. This diagram shows the quantity seven hundred eight thousand sixty-five. (Some people would call it seven hundred *and* eight thousand *and* sixty-five.) By convention, you can place a comma or space after every third digit as you go from right to left in a multi-digit numeral such as this one. Once you get to a certain nonzero digit as you work your way from right to left, all the digits farther to the left are understood to be ciphers, even though you won't usually include them or even think about them.

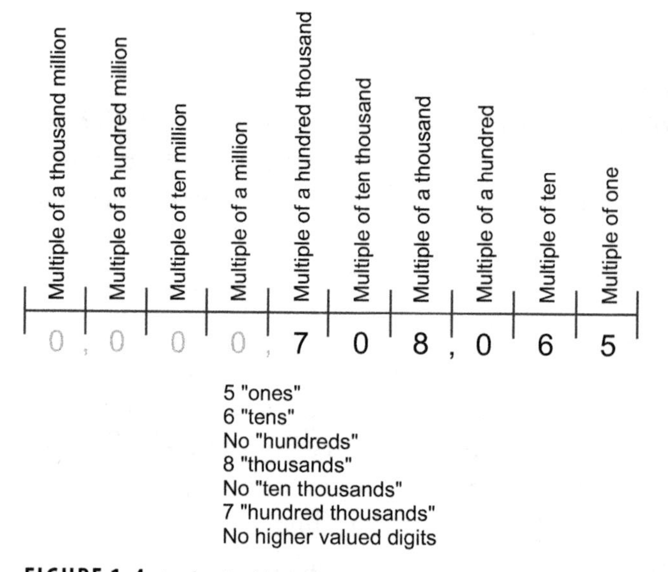

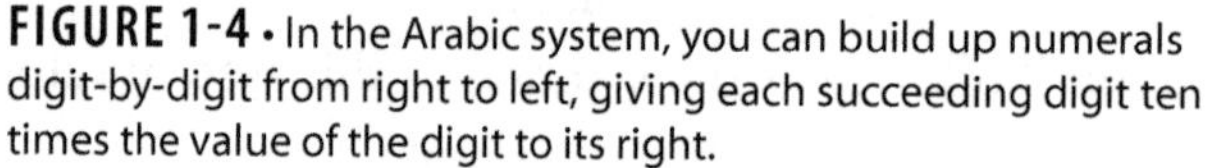
FIGURE 1-4 • In the Arabic system, you can build up numerals digit-by-digit from right to left, giving each succeeding digit ten times the value of the digit to its right.

TIP ***Every digit 0 within a multi-digit numeral clarifies the values of digits to its left. In most situations, ciphers to the left of the leftmost nonzero digit have no significance. You'll rarely see any of them written down. (How often, for example, do you find the numeral 41,567 written as 041,567 or 0,041,567?) Nevertheless, once in awhile you'll find it helpful to insert one or more of these extra ciphers during a calculation, especially a long column of numbers that you want to add up. The additional zeros can help you keep track of which column is which.***

Still Struggling

Let's make sure that we understand the difference between a *counting number* and a *whole number*. Usage varies depending on which text you consult. For our purposes, the counting numbers go as one, two, three, four, five, and so on. We can define all the counting numbers with the Roman numeration system. We'll define the whole numbers as zero, one, two, three, four, five, and so on, so the whole numbers comprise all of the counting numbers along with zero.

People who used Roman numerals rarely had to work with numbers much larger than a thousand. But in today's world of extremes, we deal with quantities that make a thousand seem tiny. Here are some of the names for numbers that we can denote by writing a numeral 1 followed by multiples of three ciphers, using the so-called *short scale* favored in the United States:

- The numeral 1 followed by three ciphers represents a *thousand*
- The numeral 1 followed by six ciphers represents a *million*
- The numeral 1 followed by nine ciphers represents a *billion*
- The numeral 1 followed by twelve ciphers represents a *trillion*
- The numeral 1 followed by fifteen ciphers represents a *quadrillion*
- The numeral 1 followed by eighteen ciphers represents a *quintillion*
- The numeral 1 followed by twenty-one ciphers represents a *sextillion*
- The numeral 1 followed by twenty-four ciphers represents a *septillion*
- The numeral 1 followed by twenty-seven ciphers represents an *octillion*

- The numeral 1 followed by thirty ciphers represents a *nonillion*
- The numeral 1 followed by thirty-three ciphers represents a *decillion*

Envision an endless string of ciphers continuing off to the left in Fig. 1-4, all of them gray (reminding you that each extra cipher stands ready to "morph" into some other digit if you want to express a huge number). If you travel to the left of the digit 7 in Fig. 1-4 by dozens of places, passing through 0 after 0, and then change one of those ciphers to the digit 1, the value of the represented number increases to a fantastic extent, demonstrating the power of the Arabic numeration system. A change in just one numeral can make a huge difference in the value of the whole number.

In the Arabic system, no limit exists as to how large a numeral you can represent. Even if a string of digits measures hundreds of kilometers long, even if it circles the earth, even if it goes from the earth to the moon—you can put a nonzero digit on the left or any digit on the right, and you get the representation for a larger whole number. Mathematicians use the term *finite* to describe anything that ends somewhere. No matter how large a whole number you want to express, the Arabic system lets you do it in a finite number of digits, and every single one of those digits comes from the basic collection {0, 1, 2, 3, 4, 5, 6, 7, 8, 9}. You needn't keep inventing new symbols when numbers get arbitrarily large, as people had to do when the Romans ruled.

TIP *You can portray every imaginable number as an Arabic numeral that contains a finite number of digits. But there's no limit to the number of whole numbers you can denote that way. Mathematicians say that the group, or set, of all whole numbers has* **infinite** *(not finite) size.*

Still Struggling

Do you still wonder why you need the digit 0? After all, it represents "nothing." Why bother with commas or spaces, either? As a matter of technical fact, you *don't* absolutely need the digit 0 and the comma in order to write legitimate Arabic numerals. The original inventors of the system put down a dot or a tiny circle instead of the full-size digit 0. However, the cipher and the comma (or space) make errors less likely when you want to do arithmetic.

PROBLEM 1-5

Imagine a whole number represented by a certain string of digits in the Arabic system. How can you change the Arabic numeral to make the number a hundred times as large, no matter what the digits are?

SOLUTION

You can make any counting numeral stand for a number a hundred times as large by attaching two ciphers to its right-hand end. Try it with a few numerals and see. Don't forget to include the commas where they belong! For example:

- **700 represents a quantity that's a hundred times as large as 7**
- **14,000 represents a quantity that's a hundred times as large as 140**
- **789,000 represents a quantity that's a hundred times as large as 7890**
- **1,400,000 represents a quantity that's a hundred times as large as 14,000**

What's the Base?

The *radix* or *base* of a numeration system is the number of single-digit symbols that we need in order to render all the counting numbers. The *radix-ten* system, also called *base-ten* or the *decimal numeration system*, therefore, has ten symbols, not counting commas (or decimal points, which we'll get into later). But we'll occasionally encounter numeration systems that use bases other than ten, and that have more or less than ten symbols to represent the digits.

Doesn't the numeral 5 always mean the quantity five, 8 always mean the quantity eight, 10 always mean the quantity ten, and 16 always mean the quantity sixteen? Not necessarily! These statements hold true in base-ten, but not necessarily in other bases. If we want to ensure that we have no ambiguity, so that we don't have to refer to any number base at all, we can write down symbols in rows. For example:

- Here are five pound signs: #####
- Here are eight pound signs: ########
- Here are ten pound signs: ##########
- Here are twelve pound signs: ############
- Here are sixteen pound signs: ################

In a *base-eight numeration system*, we'd denote the number of pound signs in the first line of the above list as the numeral 5, the number of pound signs in the second line as the numeral 10, the number of pound signs in the third line as the numeral 12, the number of pound signs in the fourth line as the numeral 14, and the number of pound signs in the last line as the numeral 20.

TIP *As you count upward from zero in the base-ten system, imagine proceeding clockwise around the face of a "ten-hour clock" as shown in Fig. 1-5A. When you've completed the first revolution, place a digit 1 to the left of the 0 and then go around again, keeping the 1 in the tens place. When you've completed the second revolution, change the tens digit to 2. You can keep going this way until you've completed the tenth revolution in which you have a 9 in the tens place. Then you must change the tens digit back to 0 and place a 1 in the hundreds place.*

TIP *The Roman numeration scheme can function as a base-five system, at least when you start counting in it. Imagine a "five-hour clock," such as the one shown in Fig. 1-5B. You start with I (which stands for the number one), not with 0. You can complete one revolution and go through part of the second, and the system works okay because after the first revolution you can keep the V and start adding symbols to its right: VI, VII, VIII. But when you get past VIII (which stands for the number eight), you run into a problem. The number nine is not represented as VIV, although technically it could be. It's written as IX, but X does not appear anywhere*

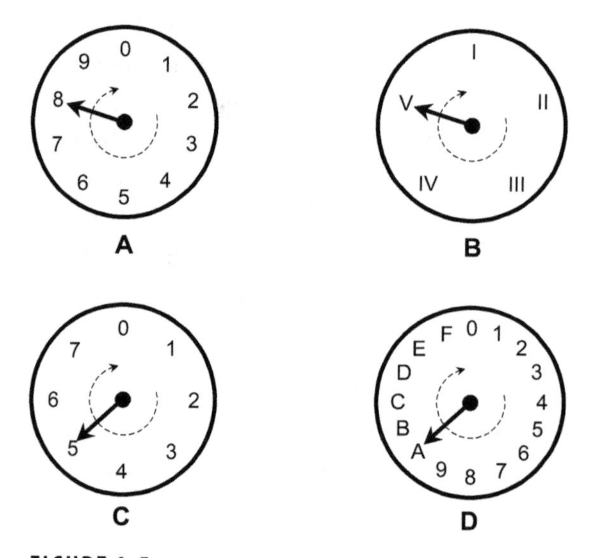

FIGURE 1-5 • Clock-like representations of digits in base-ten or decimal (A), Roman base-five (B), base-eight or octal (C), and base-sixteen or hexadecimal (D). (The octal and hexadecimal systems are discussed on pages 13 through 16.)

on the clock face. The orderliness of this system falls apart before you even get twice around!

PROBLEM 1-6

Refer to the list of pound-sign quantities on page 11. If we work in a base-four numeration system, how should we denote the quantities in each line?

SOLUTION

In the base-four scheme, we have only the four numerals 0, 1, 2, and 3 available. Let's move in each numeral from right to left. We denote the number of pound signs in the first line as the numeral 11 (one times one, plus one times four), the number of pound signs in the second line as the numeral 20 (zero times one, plus two times four), the number of pound signs in the third line as the numeral 22 (two times one, plus two times four), the number of pound signs in the fourth line as the numeral 30 (zero times one, plus three times four), and the number of pound signs in the last line as the numeral 40 (zero times one plus four times four).

PROBLEM 1-7

Refer to the list of pound-sign quantities one more time! If we work in a base-six numeration system, how should we denote the quantities in each line?

SOLUTION

In the base-six scheme, we have the six numerals 0, 1, 2, 3, 4, and 5 available. Once again, we move in each numeral from right to left. We denote the number of pound signs in the first line as the numeral 5 (five times one), the number of pound signs in the second line as the numeral 12 (two times one, plus one times six), the number of pound signs in the third line as the numeral 14 (four times one, plus one times six), the number of pound signs in the fourth line as the numeral 20 (zero times one, plus two times six), and the number of pound signs in the last line as the numeral 24 (four times one, plus two times six).

Base Eight

Figure 1-5C shows an "eight-hour clock" that can demonstrate how the base-eight or *octal numeration system* works. You can use the same upward-counting scheme as you did with the "ten-hour clock," skipping the digits 8 and 9 (which don't exist in the octal system). When you finish the first revolution and you're ready to start the second, place a 1 to the left of the digits shown, so you count

$$\cdots 5, 6, 7, 10, 11, 12 \cdots$$

The string of three dots, called an *ellipsis*, indicates that a pattern continues for awhile, or perhaps even forever, saving you from having to do a lot of symbol scribbling. (You'll see this notation often in mathematics.) Continuing through the second revolution and into the third, you count

$$\cdots 15, 16, 17, 20, 21, 22 \cdots$$

When you finish up the eighth revolution and enter the ninth, you count

$$\cdots 75, 76, 77, 100, 101, 102 \cdots$$

PROBLEM 1-8

Convert the octal numeral 123 to its equivalent in the decimal system.

SOLUTION

In the octal system, the digit farthest to the right represents the multiple of the decimal 1. As we go to the left:

- **The second digit from the right tells us the multiple of the decimal 8**
- **The third digit from the right tells us the multiple of the decimal 8×8, or 64**
- **The fourth digit from the right tells us the multiple of the decimal $8 \times 8 \times 8$, or 512**
- **The fifth digit from the right tells us the multiple of the decimal $8 \times 8 \times 8 \times 8$, or 4096**

And so on it goes, as far to the left as we want! If we see the octal numeral 123, we know that it means 1×64 plus 2×8 plus 3×1 in decimal form. When we add those numbers up, we get $64 + 16 + 3 = 83$.

Base Sixteen

Let's invent one more strange clock. It has 16 hours as shown in Fig. 1-5D. You can see from this drawing how the base-16 or *hexadecimal numeration system* works. Use the same upward-counting scheme as you did with the "ten-hour clock" and the "eight-hour clock." The hexadecimal system contains some new digits, in addition to those in the base-ten system:

- A stands for the decimal quantity ten
- B stands for the decimal quantity eleven
- C stands for the decimal quantity twelve
- D stands for the decimal quantity thirteen
- E stands for the decimal quantity fourteen
- F stands for the decimal quantity fifteen

When you finish the first revolution and move into the second, place a 1 to the left of the digits shown. You count

$$\cdots 8, 9, A, B, C, D, E, F, 10, 11, 12, 13, \cdots$$

Continuing through the second revolution and into the third, you count

$$\cdots 18, 19, 1A, 1B, 1C, 1D, 1E, 1F, 20, 21, 22, 23, \cdots$$

When you complete the tenth revolution and move into the eleventh, you count

$$\cdots 98, 99, 9A, 9B, 9C, 9D, 9E, 9F, A0, A1, A2, A3, \cdots$$

It goes on like this with B, C, D, E, and F in the 16s place. Then you get to the end of the sixteenth revolution and move into the seventeenth:

$$\cdots F8, F9, FA, FB, FC, FD, FE, FF, 100, 101, 102, 103, \cdots$$

PROBLEM 1-9

Convert the hexadecimal numeral 2D03 to decimal form. Don't use a computer or go on the Internet to find a Web site that will do it for you. Do it the "long way"!

SOLUTION

To solve this problem, you must know the place values. The digit farthest to the right represents a multiple of the decimal 1. The next digit to the left represents a multiple of the decimal quantity sixteen. After that comes a multiple of the decimal two hundred fifty-six (or 16×16). Then comes a multiple of the decimal 4096 (or $16 \times 16 \times 16$). Note that the hexadecimal D represents the decimal quantity thirteen. Now you can figure out the conversion as follows:

- **In the 1s place you have 3, so you start out with 3**
- **In the 16s place you have 0, so you add $0 \times 16 = 0$ to what you have so far**
- **In the 256s place you have D, so you add $13 \times 256 = 3328$ to what you have so far**

- **In the 4096s place you have 2, so you add $2 \times 4096 = 8192$ to what you have so far**

Because no digits appear to the left of 2 in the hexadecimal expression, you're finished at this point. The final result, expressed as a sum in decimal numerals, is

$$3 + 0 + 3328 + 8192 = 11{,}523$$

Base Two

When engineers began to build computers in the twentieth century, they wanted to denote numbers using only two digits: 0 to represent the "off" condition of an electrical switch, and 1 to represent the "on" condition. We can also call these states false/true, no/yes, low/high, or negative/positive, giving a *binary numeration system*. Figure 1-6 shows how we assemble numerals in the binary system. We simply double the value of each digit as we move place-by-place to the left. We start with a ones place at the extreme right, then a decimal twos place to the left of that, a decimal fours place to the left of that, a decimal eights place to the left of that, and so on.

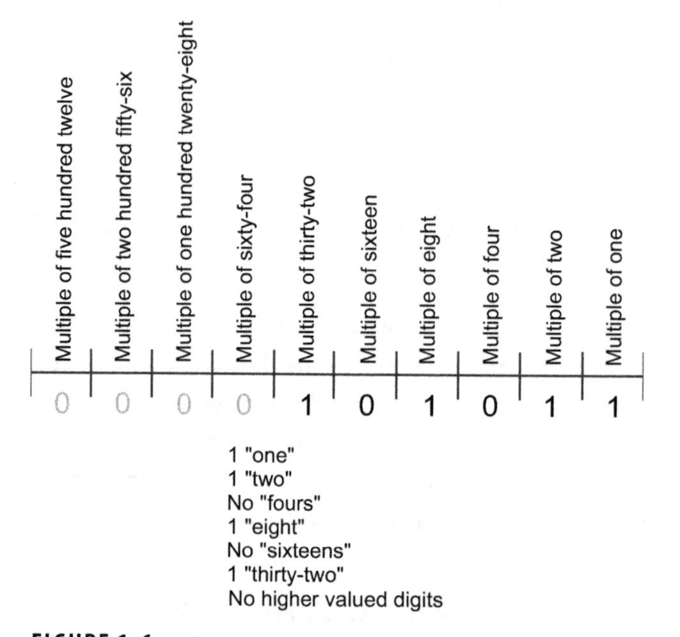

FIGURE 1-6 • In the binary system, you can build up numerals digit-by-digit from right to left, giving each succeeding digit twice the value of the digit to its right. Note the absence of commas in this system.

TIP *Every binary numeral has a unique equivalent in the decimal system, and vice-versa. When you use a computer or calculator and input a sequence of decimal digits, the machine converts it into a binary numeral, performs whatever calculations or operations you demand, converts the result back to a decimal numeral, and then displays that numeral. All of the conversions, calculations, and switching actions take place out of sight, at incredible speed.*

Still Struggling

Table 1-2 compares numerical values in the decimal, binary, octal, and hexadecimal systems from the decimal quantity zero to the decimal quantity sixty-four. Using this table, you can figure out (with a little thought and scribbling) how to convert larger decimal numerals to any of the other forms. Fortunately, the Internet offers plenty of computer programs and Web sites that will do such conversions for you!

PROBLEM 1-10

Convert the binary numeral 10101 to decimal form.

SOLUTION

In a binary expression, the digit farthest to the right represents a multiple of the decimal 1. The next digit to the left represents a multiple of the decimal two. After that comes a multiple of the decimal four, then a multiple of the decimal eight, then a multiple of the decimal sixteen, and so on. You can break the conversion down as follows:

- **In the ones place you have 1, so you start out with 1**
- **In the twos place you have 0, so you add $0 \times 2 = 0$ to what you have so far**
- **In the fours place you have 1, so you add $1 \times 4 = 4$ to what you have so far**
- **In the eights place you have 0, so you add $0 \times 8 = 0$ to what you have so far**
- **In the sixteens place you have 1, so you add $1 \times 16 = 16$ to what you have so far**

That's as far as you can go. The final result, expressed as a sum in decimal numerals, works out as

$$1 + 0 + 4 + 0 + 16 = 21$$

TABLE 1-2 The conventional (or decimal) numerals 0 through 64, along with their binary, octal, and hexadecimal equivalents.

Decimal	Binary	Octal	Hexadecimal
0	0	0	0
1	1	1	1
2	10	2	2
3	11	3	3
4	100	4	4
5	101	5	5
6	110	6	6
7	111	7	7
8	1000	10	8
9	1001	11	9
10	1010	12	A
11	1011	13	B
12	1100	14	C
13	1101	15	D
14	1110	16	E
15	1111	17	F
16	10000	20	10
17	10001	21	11
18	10010	22	12
19	10011	23	13
20	10100	24	14
21	10101	25	15
22	10110	26	16
23	10111	27	17
24	11000	30	18
25	11001	31	19
26	11010	32	1A
27	11011	33	1B
28	11100	34	1C
29	11101	35	1D
30	11110	36	1E
31	11111	37	1F

TABLE 1-2 The conventional (or decimal) numerals 0 through 64, along with their binary, octal, and hexadecimal equivalents. (*Continued*)

Decimal	Binary	Octal	Hexadecimal
32	100000	40	20
33	100001	41	21
34	100010	42	22
35	100011	43	23
36	100100	44	24
37	100101	45	25
38	100110	46	26
39	100111	47	27
40	101000	50	28
41	101001	51	29
42	101010	52	2A
43	101011	53	2B
44	101100	54	2C
45	101101	55	2D
46	101110	56	2E
47	101111	57	2F
48	110000	60	30
49	110001	61	31
50	110010	62	32
51	110011	63	33
52	110100	64	34
53	110101	65	35
54	110110	66	36
55	110111	67	37
56	111000	70	38
57	111001	71	39
58	111010	72	3A
59	111011	73	3B
60	111100	74	3C
61	111101	75	3D
62	111110	76	3E
63	111111	77	3F
64	1000000	100	40

QUIZ

Refer to the text in this chapter if necessary. A good score is eight correct. Answers are in the back of the book.

1. **Imagine that you have an unlimited supply of sugar cubes, all of which measure exactly one centimeter (a hundredth of a meter) along each edge. If you arrange a bunch of these little things into a hundred-by-hundred-cube array, thereby getting a large, flat "sugar slab" measuring one meter wide by one meter deep by one centimeter tall, how many small sugar cubes do you have in total? (Assume that you work in the decimal system.)**
   A. 10,000,000
   B. 1,000,000
   C. 100,000
   D. 10,000

2. **If you neatly stack a great many of the little sugar cubes described above into a hundred-by-hundred-by hundred-cube mass, thereby getting a "sugar block" measuring one meter wide by one meter deep by one meter tall, how many sugar cubes do you have in total? (Assume that you work in the decimal system.)**
   A. 10,000,000
   B. 1,000,000
   C. 100,000
   D. 10,000

3. **In the decimal system, the Roman numeral MCXI represents**
   A. 1111.
   B. 1911.
   C. 2005.
   D. 1956.

4. **In the octal numeral 5103, the digit 5 represents the decimal quantity**
   A. 2560.
   B. 1280.
   C. 640.
   D. 320.

5. **In Roman numerals, we represent the decimal 52 as**
   A. XII.
   B. IIX.
   C. LII.
   D. LVV.

6. **In the decimal numeral 335,427, the digit 4 represents**
   A. 40.
   B. 400.
   C. 4000.
   D. 40,000.

7. In the octal numeration system, what follows 67?

A. 70
B. 68
C. 80
D. 100

8. In binary numerals, we represent the Roman VIII as

A. 1110.
B. 1000.
C. 1101.
D. 1011.

9. In the binary numeral 101010101, the leftmost digit represents

A. 64.
B. 128.
C. 256.
D. 512.

10. In the Roman numeration system, what symbol represents zero?

A. O
B. N
C. Z
D. The Roman system has no symbol for zero.

chapter 2

Quantities

Over the centuries, numbers have evolved from counting aids to scientific tools. Numbers have given rise to philosophers' frustrations and geeks' games. Most of us use numbers to tell us "how many" of this or "how much" of the other thing, and don't dwell on them much beyond that. But numbers have interesting properties in their own right.

CHAPTER OBJECTIVES

In this chapter, you will

- See how numbers work for counting.
- Compare even and odd numbers.
- Define and identify prime numbers.
- Break whole numbers into products of primes.
- Discover negative numbers.
- Define and identify integers.
- Combine integers to get fractions.

Counting and Whole Numbers

We build up the *set* (collection) of *counting numbers*, also called the *natural numbers*, from a starting point of 1. We can write down the set of counting numbers as an "endless list" enclosed in curly brackets (technically called *braces*) as follows:

$$\{1, 2, 3, \cdots, n, \cdots\}$$

where n represents any counting number we want. When we add 0 to the set of counting numbers, we get the set of *whole numbers*, which we can write out by adding 0 at the beginning of the "endless list" of counting numbers:

$$\{0, 1, 2, 3, 4, \cdots, n, \cdots\}$$

Once again, n represents any whole number we want. We can visualize the counting numbers or the whole numbers as points evenly spaced along a *half-line* (also called a *ray*), where the size of the number varies in direct proportion to its distance from the starting point. Figure 2-1 shows a *whole-number half-line*. We can make it into a *counting-number half-line* if we remove the point for 0.

TIP ***The whole-number half-line and the counting-number half-line both contain infinitely many points. Although the whole-number half-line contains one more point than the counting-number half-line does, the two lines have equally many points! (If you add one to "infinity," or if you take away one from "infinity," you still end up with "infinity.")***

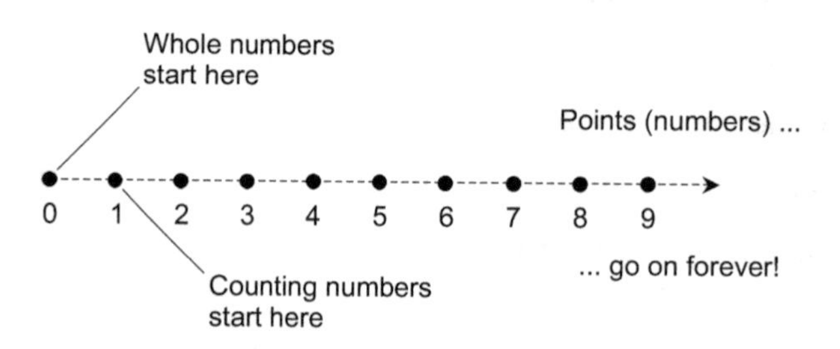

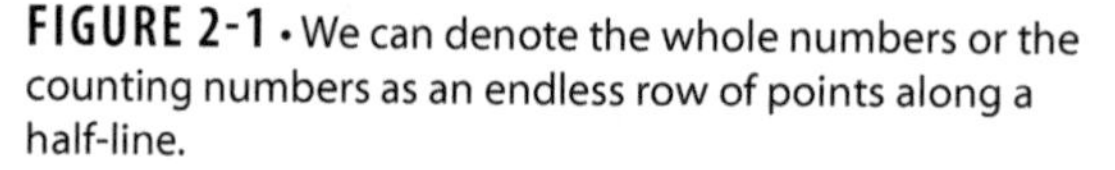

FIGURE 2-1 • We can denote the whole numbers or the counting numbers as an endless row of points along a half-line.

Still Struggling

Diverse authors have used the terms *whole number*, *natural number*, and *counting number* interchangeably, and have never reached a total consensus as to whether or not they intend to include 0 in any of the sets. This indecision dates back more than a century! If you want to make sure that your readers will never get confused, you can call the whole numbers as defined here (including 0) the *nonnegative whole numbers*, and you can call the counting or natural numbers as defined here (not including 0) the *positive whole numbers*.

PROBLEM 2-1

How can we explain in formal terms, and without resorting to vague expressions such as "infinity," why equally many whole numbers and counting numbers exist?

SOLUTION

Whenever a mathematician wants to prove that two sets contain the same quantity of objects, she finds a way to compare the sets "side-by-side" so that each object in one set has exactly one mate in the other set. Once she's solved that problem, she's found a *one-to-one correspondence* between the objects in the sets. In the case of the whole numbers versus the counting numbers, we can pair off each whole number with its counting-number mate by adding 1 to the whole number. Conversely, we can pair off each counting number with its whole-number mate by subtracting 1 from the counting number.

PROBLEM 2-2

Draw a diagram with two half-lines side-by-side, illustrating how the one-to-one correspondence works in the solution to Problem 2-1.

SOLUTION

Figure 2-2 shows the whole-number half-line (top) and the counting-number half-line (bottom) running parallel to each other with their starting

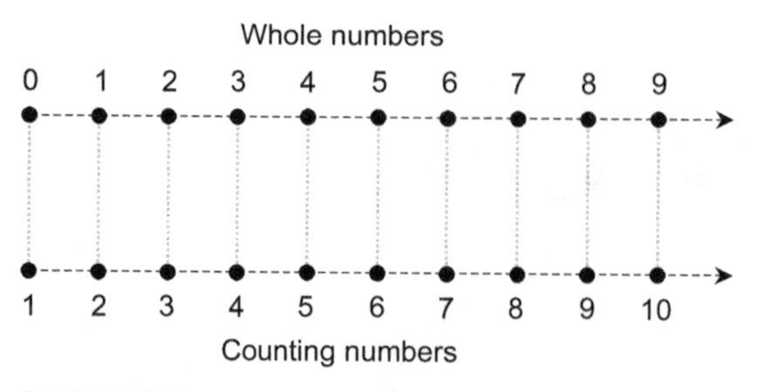

FIGURE 2-2 • Illustration for Problem 2-2 and its solution.

points aligned. On both half-lines, the point sequences continue to the right without end. We can pair up the points one-for-one as shown by the vertical, dotted gray lines, no matter how far out along either half-line we go.

Even and Odd Whole Numbers

Imagine that we count up through the whole numbers starting with 0 and then skipping every other number. We write the numerals down to get the list

$$\{0, 2, 4, 6, 8, 10, 12, \cdots\}$$

We've denoted the set of *even whole numbers*. Figure 2-3 shows the even values as light gray dots along the whole-number half-line.

Now let's start over, but this time, we'll begin at 1 instead of 0, and skip every other number to get the list

$$\{1, 3, 5, 7, 9, 11, 13, \cdots\}$$

That's the set of *odd whole numbers*. In Fig. 2-3, the odd values correspond to the black dots along the whole-number half-line.

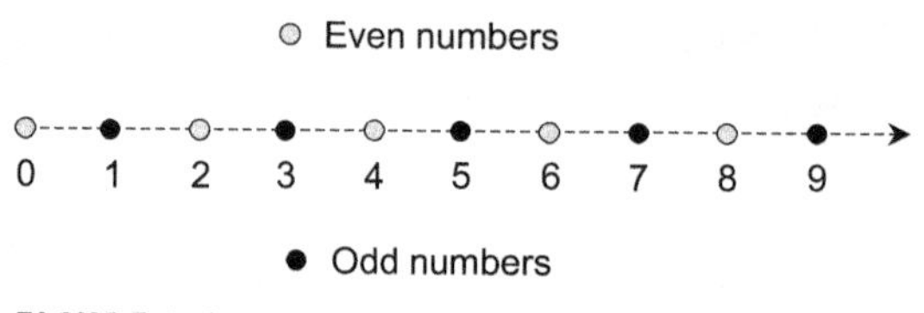

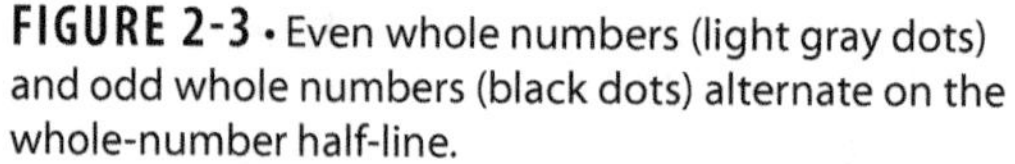

FIGURE 2-3 • Even whole numbers (light gray dots) and odd whole numbers (black dots) alternate on the whole-number half-line.

TIP *By definition, if you cut an even number in half, you always get a whole number. Also by definition, the odd whole numbers are precisely those that aren't even, so if you cut an odd number in half, you never get a whole number.*

TIP *If you double a whole number, regardless of whether it's even or odd, you always end up with an even number.*

PROBLEM 2-3

Equally many even numbers and whole numbers exist. How can we demonstrate the reason why?

SOLUTION

We can pair off the even numbers and the whole numbers in a one-to-one correspondence. When we want to find the whole-number mate of any particular even number, we cut the even number in half. For example:

- 3 is the whole-number mate of the even number 6
- 10 is the whole-number mate of the even number 20
- 17 is the whole-number mate of the even number 34

When we want to find the even-number mate of any particular whole number, we double the whole number. For example:

- 10 is the even-number mate of the whole number 5
- 24 is the even-number mate of the whole number 12
- 46 is the even-number mate of the whole number 23

PROBLEM 2-4

Draw a diagram with two half-lines side-by-side to illustrate the correspondence we described in the solution to Problem 2-3.

SOLUTION

Figure 2-4 portrays, in graphical form, how the one-to-one correspondence works between the even and whole numbers as described in the solution to Problem 2-3. The vertical, dotted gray lines show the individual correspondences.

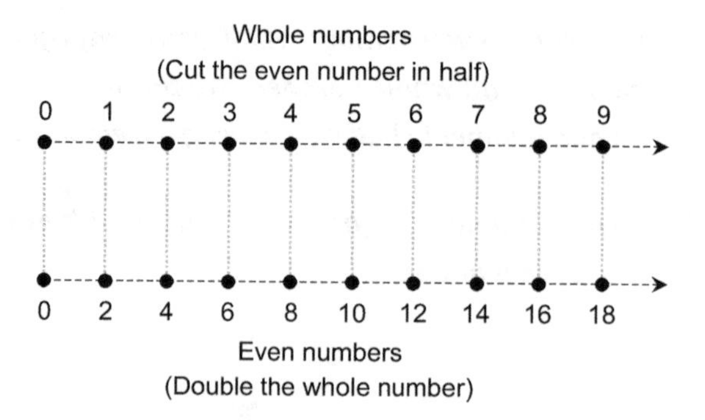

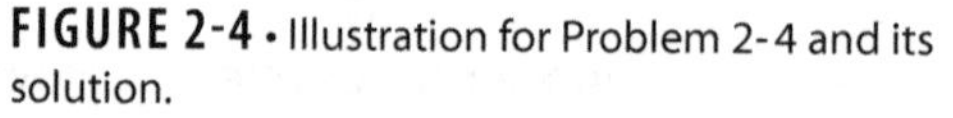
FIGURE 2-4 • Illustration for Problem 2-4 and its solution.

PROBLEM 2-5

How can we explain why equally many odd numbers and whole numbers exist?

SOLUTION

We can pair off the odd and whole numbers in a one-to-one correspondence. To determine the whole-number mate of any particular odd number, we subtract 1 from the odd number and then cut the result in half. For example:

- 2 is the whole-number mate of the odd number 5
- 8 is the whole-number mate of the odd number 17
- 15 is the whole-number mate of the odd number 31

To calculate the odd-number mate of any particular whole number, we double the whole number and then add 1 to the result. For example:

- 7 is the odd-number mate of the whole number 3
- 13 is the odd-number mate of the whole number 6
- 59 is the odd-number mate of the whole number 29

PROBLEM 2-6

Draw a diagram with two half-lines side-by-side to graphically portray the one-to-one correspondence described in the solution to Problem 2-5.

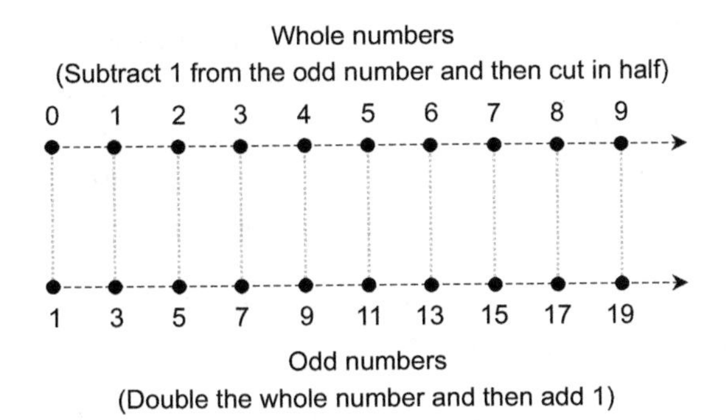

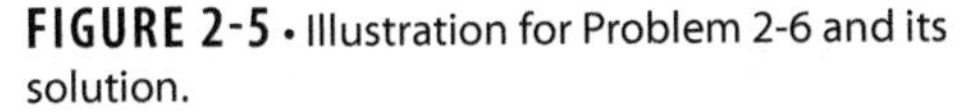
FIGURE 2-5 • Illustration for Problem 2-6 and its solution.

SOLUTION

Figure 2-5 shows how the one-to-one correspondence works between the odd and whole numbers, as described in the solution to Problem 2-5. The vertical, dotted gray lines show the individual correspondences.

Prime Numbers

We can break any whole number down into a product of whole numbers called *factors*. Sometimes we can carry out this process in only one way; more often we can do it in many ways. For example, we can split up the whole number 100 as follows:

$$100 = 100 \times 1$$

$$100 = 10 \times 10$$

$$100 = 20 \times 5$$

$$100 = 25 \times 4$$

$$100 = 50 \times 2$$

$$100 = 10 \times 5 \times 2$$

$$100 = 5 \times 5 \times 4$$

$$100 = 5 \times 5 \times 2 \times 2$$

As we take larger and larger whole numbers, we can *usually* find more and more different products of factors that "multiply up to it." But exceptions occur! For example, if we try to break 101 down into whole-number factors, we find that we can do it in only one way: 101×1. The same thing happens with a lot of other numbers (infinitely many, in fact). If we encounter a whole number larger than 1 and we can't split it into factors other than itself and 1, then we call it a *prime number*, or simply a *prime*.

TIP ***Most mathematicians don't regard 1 as a prime number, even though its only whole-number factors are itself and 1.***

TIP ***Whenever we come across a prime number, no matter how large, we can always find another prime number that's larger. However, if we try to "predict" the value of that next-larger prime number, we'll find the task impossible. We must test a whole number for factors to figure out whether or not it's a prime. As you might guess, a computer comes in handy for that purpose!***

Still Struggling

Table 2-1 lists the first 50 prime numbers. Can you see a pattern? I can't! As far as I know, no pattern exists. The prime numbers seem "randomly scattered" throughout the boundless expanse of the whole numbers. In any case, we know one thing for sure: However long we keep "grinding out" larger and larger primes, we'll never finish the list, even if we live as long as the earth does. Mathematicians have proven that there are infinitely many prime numbers.

TABLE 2-1 The first 50 prime numbers, each of which we can split into only two different factors: itself and 1.

2	31	73	127	179
3	37	79	131	181
5	41	83	137	191
7	43	89	139	193
11	47	97	149	197
13	53	101	151	199
17	59	103	157	211
19	61	107	163	223
23	67	109	167	227
29	71	113	173	229

PROBLEM 2-7

How many *even* prime numbers exist?

SOLUTION

Only one! According to the foregoing definition, 2 is the only even prime. Whenever we take any even number larger than 2, we can break it down into 2 times some whole number that's smaller than itself but larger than 1. For example:

$$4 = 2 \times 2$$

$$6 = 3 \times 2$$

$$8 = 4 \times 2$$

$$10 = 5 \times 2$$

$$12 = 6 \times 2$$

$$\downarrow$$

and so on, forever

Prime Factors

Prime numbers have a property that makes them ideal for finding factors of whole numbers: We can break down any nonprime whole number into a product of primes, which we call *prime factors*. For this reason, mathematicians call any product of primes a *composite number*. All the composites are *composed* of primes.

When we want to find the prime factors of a large number, we start by using a calculator to find the *square root* of the number. On my computer's calculator, I can find a square root by entering the desired number and then clicking on a button marked with the *radical sign* (√). Once we've found the square root, we divide the original number by all the primes less than or equal to that square root. If we ever get a whole-number quotient during this process, then we know that the divisor and the quotient are both factors of the original number. Sometimes the quotient turns out prime, and sometimes it doesn't. If it doesn't, then we can factor it down further into primes. We must keep dividing the original by progressively smaller primes until we get down to 2.

Usually, the square root of a whole number is *not* another whole number. We need not worry about that detail when we want to find prime factors. We simply round the square root up to the next larger whole number, and then look up all the primes less than or equal to it.

TIP *If Table 2-1 doesn't extend far enough when you want to factor a large whole number into a product of primes, you can find plenty of prime-number lists on the Internet, some of which go beyond anything you'll ever need.*

PROBLEM 2-8

Find the prime factors of 139.

SOLUTION

When you use a calculator to find the square root of 139, you'll get a number between 11 and 12. Round it up to 12. Now find all the prime numbers less than or equal to 12. From Table 2-1, you can see that they're 2, 3, 5, 7, and 11. Divide 139 by each of these primes and look for whole-number quotients, as follows:

$$139/11 = 12.\#\#\# \cdots \text{ (Not whole)}$$

$$139/7 = 19.\#\#\# \cdots \text{ (Not whole)}$$

$$139/5 = 27.8 \text{ (Not whole)}$$

$$139/3 = 46.333 \cdots \text{ (Not whole)}$$

$$139/2 = 69.5 \text{ (Not whole)}$$

Here, "###" means a string of numbers with a recurring pattern too long to write down easily. (The pattern doesn't matter if the number isn't whole.) You know from the foregoing calculations that the only factors of 139 are 1 and itself, so by definition, 139 is prime.

PROBLEM 2-9

Find the prime factors of 493.

SOLUTION

Using a calculator to find the square root and then going up to the next higher whole number, you get 23. From Table 2-1, you can see that the

primes that do not exceed 23 are 2, 3, 5, 7, 11, 13, 17, 19, and 23. Now divide 493 by each of these quantities and look for whole numbers:

$$493/23 = 21.\#\#\# \cdots \text{ (Not whole)}$$

$$493/19 = 25.\#\#\# \cdots \text{ (Not whole)}$$

$$493/17 = 29 \text{ (Whole)}$$

$$493/13 = 37.\#\#\# \cdots \text{ (Not whole)}$$

$$493/11 = 44.818181 \cdots \text{ (Not whole)}$$

$$493/7 = 70.\#\#\# \cdots \text{ (Not whole)}$$

$$493/5 = 98.6 \text{ (Not whole)}$$

$$493/3 = 164.333 \cdots \text{ (Not whole)}$$

$$493/2 = 246.5 \text{ (Not whole)}$$

The prime factors of 493 are 17 and 29, because 17 and 29 both appear in Table 2-1. You can't factor 493 down in any other way, except, of course, as 493×1.

Negative Numbers and Integers

What do mathematicians and scientists mean when they talk about *negative numbers*? That question runs deeper than it seems at first thought. We can partially answer it, in practical terms, by imagining situations where we find negative numbers useful.

In the United States, most nonscientific people use the *Fahrenheit temperature scale*, where 32 degrees represents the freezing point of water. Scientists, and people outside the United States, use the *Celsius temperature scale*, where 0 degrees represents the freezing point of water. In either system, temperatures often get colder than 0 degrees. Then people use the words "negative" or "minus" to preface temperature values. For example, "negative 10 degrees" or "minus 10 degrees" translates to a temperature that's 10 degrees colder than 0 degrees (in whatever scale we want to use). Some people would say "10 below zero" or "10 below."

Here's another real-life situation where negative numbers come in handy. These days, nearly everyone has a credit card. When you first get the card, it has a balance of 0 dollars ($0.00); you haven't deposited any money in the card's account, but you don't owe the bank any money, either. (At least that's the case if you deal with a good bank!) What if you buy some items at the local department store, "charging" up a balance of $49.00? How much money is in the account now? If you think of it as the bank's account, then they have a claim to $49.00 of your money. If you think of it as your account, then you're $49.00 in debt. You have, in a sense, negative $49.00. If you go to another store and charge $10.25 more, you'll end up with negative $59.25. In theory, there's no limit to how *large negatively* your account, in dollars, can become. (In practice, the bank will put a limit on it.)

We denote a negative whole number by putting a minus sign in front of a counting number. The exception is 0, where a negative sign doesn't change the meaning. "Negative 0" is the same thing as "positive 0" in ordinary mathematics. In the credit-card situation just described, you start out with $0.00 and then go to –$49.00, then to –$59.25. The same thing can happen with temperature. If it was 0 degrees yesterday afternoon and then the temperature fell by 10 degrees overnight, it was –10 degrees in the morning.

When we combine all the negative whole numbers with all the "conventional" whole numbers (0, 1, 2, 3, and so on), we get the *integers*, which we can portray as the following "endless list":

$$\{\cdots, -3, -2, -1, 0, 1, 2, 3, \cdots\}$$

TIP ***Figure 2-6 shows a line along which we represent the integers as points. The integers get smaller as we move to the left, and larger as we move to the right. For example, –5 is smaller (or less) than –2, and any negative integer is smaller (or less) than any counting number. Conversely, –2 is larger (or greater) than –5, and any counting number is larger (or greater) than any negative integer.***

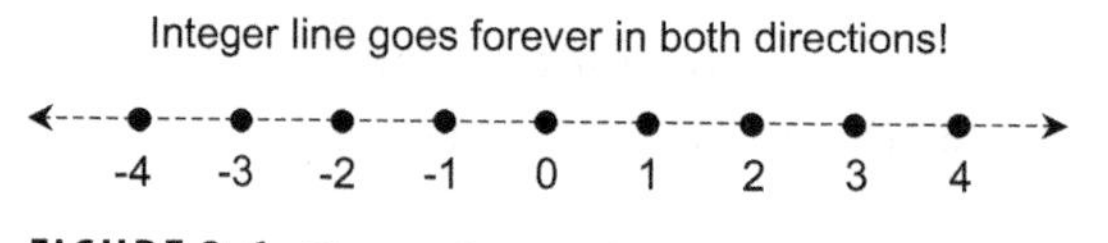

FIGURE 2-6 • We can denote the integers as an endless row of points along a line that runs infinitely far to the left and infinitely far to the right.

Still Struggling

You ask, "How can an integer such as, say, −158 be smaller than −12? If I find myself in debt by $158.00, isn't it a *bigger* problem than if I'm in debt by $12.00?" In the pure mathematical sense, the integer −158 is indeed smaller than −12, just as a temperature of −158 degrees is colder than −12 degrees. In fact, the integer −158 is less than −12 or −32 or −157. But the value of −158 is *larger negatively* than the value of −12 or −32 or −157. If you want to avoid confusion when comparing numbers, you'd better choose your words carefully!

PROBLEM 2-10

In the wild world of mathematics, equally many integers and nonnegative whole numbers exist. How can we pair off the integers with the nonnegative whole numbers in a one-to-one correspondence to demonstrate this fact?

SOLUTION

We can pair off the numbers in "parallel sequences" with the nonnegative whole numbers ascending and the integers alternating, as follows:

- **The whole number 0 maps to and from the integer 0**
- **The whole number 1 maps to and from the integer 1**
- **The whole number 2 maps to and from the integer −1**
- **The whole number 3 maps to and from the integer 2**
- **The whole number 4 maps to and from the integer −2**
- **The whole number 5 maps to and from the integer 3**
- **The whole number 6 maps to and from the integer −3**
- **The whole number 7 maps to and from the integer 4**
- **The whole number 8 maps to and from the integer −4**
- **The whole number 9 maps to and from the integer 5**

and so on, for as long as we want to keep going!

PROBLEM 2-11

Draw a diagram that shows the one-to-one correspondence described in the solution to Problem 2-10.

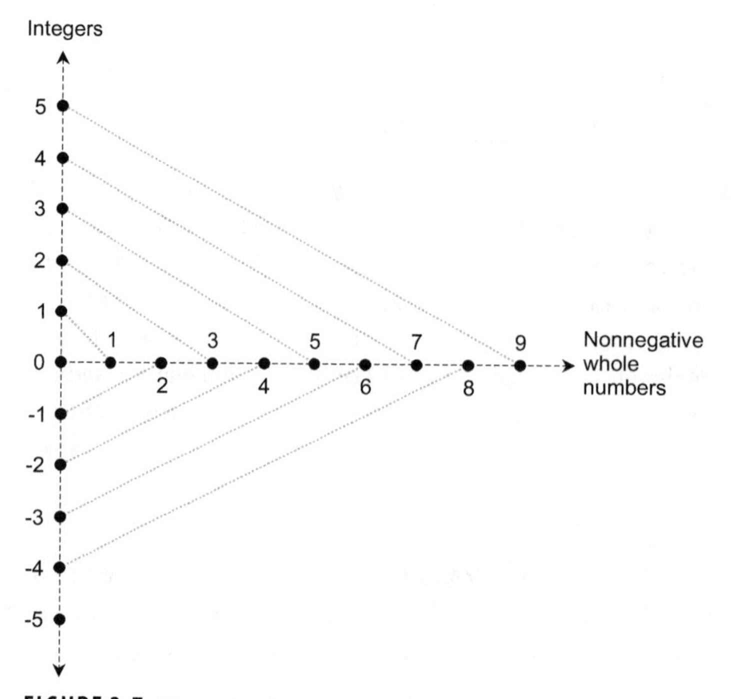

FIGURE 2-7 • Illustration for Problem 2-11 and its solution.

SOLUTION

Figure 2-7 puts the one-to-one correspondence described in the solution to Problem 2-10 into pictorial form. The slanting, dotted gray lines show the individual correspondences.

Between the Integers

Now that we've defined all the positive and negative whole numbers, let's think about *nonwhole fractions* (fractions other than those that, by happenstance, work out as whole numbers). Imagine a circular dessert pie. We want to divide the pie up equally among a certain number of people. For any particular pie, as the number of hungry people increases, each person will get a smaller portion. If we want to divide a pie equitably among three people, then we'll cut it into three equal parts called *thirds*. If we have four people, we'll cut the pie into four equal parts called *fourths*. For five people, we'll cut the pie into five equal parts called *fifths*. If we have seven people, we'll slice up the pie to get seven equal portions called *sevenths*. Using numerals and a forward slash:

- We denote one-third by writing 1/3 as shown in Fig. 2-8A
- We denote one-fourth by writing 1/4 as shown in Fig. 2-8B
- We denote one-fifth by writing 1/5 as shown in Fig. 2-8C
- We denote one-seventh by writing 1/7 as shown in Fig. 2-8D

In general, for any counting number n, we can write one-nth as $1/n$.

TIP ***As we bring more and more people to a dining-room table with a single pie, we'll have to convince each person to accept smaller and smaller portions if we want all the people to get equal amounts. As the number of people grows without limit, the size of each person's portion diminishes toward nothing! As the value of a counting number n increases without limit, the value of the fraction 1/n approaches 0.***

Imagine that our guests have received their portions of pie, but for some reason, not all of the people want to actually eat the stuff! For example, we might have three people at the table, but only two of them want pie. If we've cut up

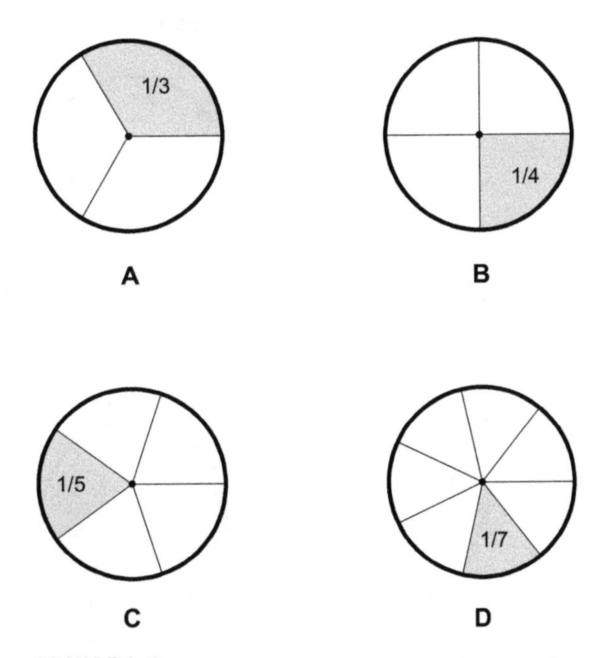

FIGURE 2-8 • We can divide an object into three equal parts to get thirds (at A), four equal parts to get fourths (at B), five equal parts to get fifths (at C), and seven equal parts to get sevenths (at D).

the pie into thirds, we'll end up serving two of those thirds (two-thirds or 2/3), leaving the remainder shown in Fig. 2-9A. If we have four people at the table and we cut up the pie into fourths and then find out that only three people want pie, then we'll end up serving three-fourths (3/4) of the pie, as shown in Fig. 2-9B. If we have five people and it turns out that only three of them hanker for pie, we'll serve three-fifths (3/5) of the dessert, as shown in Fig. 2-9C. If seven people sit at the table and only four of them pine for a piece of pie after we've cut it up, we'll give away four-sevenths (4/7) of the pie, as shown in Fig. 2-9D. Regardless of how many people sit at our table, and regardless of how many of those people want the pie that we offer, we'll always end up serving some fractional amount less than or equal to 1 (the whole pie, if everyone wants some) and greater than or equal to 0 (no pie at all, if nobody wants any).

TIP ***We've just generated, in mouth-watering terms, a scheme with which we can define all possible nonwhole fractions between and including 0 and 1.***

Nonwhole fractions exist outside the range between 0 and 1. We can get any possible nonwhole fractional quantity by adding or subtracting counting

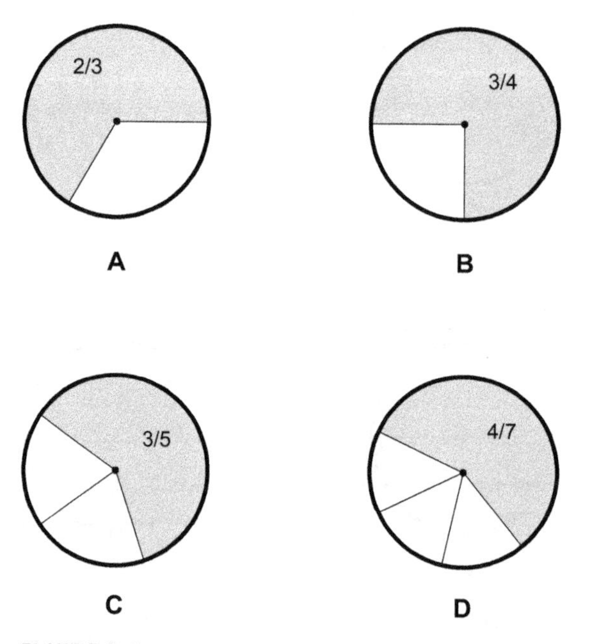

FIGURE 2-9 • Multiple pieces of an object give us fractional parts such as two-thirds (at A), three-fourths (at B), three-fifths (at C), and four-sevenths (at D).

numbers from a fraction that lies between 0 and 1. For example, when we add 1 to 2/3, we get 1-2/3. (Here, the dash does *not* act a minus sign! Instead, it visually separates the whole-number part of the numeral from the nonwhole fractional part.) We can also call the foregoing quantity 5/3, because 1 equals 3/3, and when we add 2/3 to 3/3 we get 5/3.

Now imagine a situation in which we expect a large number of guests to visit our dessert table. We decide that one pie won't provide enough for everybody, so we buy three pies. We decide that 1/7 of a pie constitutes a reasonable portion for each person: enough to satisfy, but not so much as to overfill! Suppose that 15 hungry people show up, and they all promise to eat their pie. We cut each of the three pies into sevenths, and give each of person one piece. Figure 2-10 shows the result. We account for two whole pies plus 1/7 of the third pie. (We keep the remaining 6/7 of the third pie for our own family, so that we can eat it at a later date, like tomorrow!) We've served 2-1/7 pies to our dessert guests. Alternatively, we can say that we've given away 15/7 pies.

We can find fractions having infinitely many values greater than or equal to 0 when we divide a whole number by a counting number. Usually, such fractions will turn out nonwhole, such as 15/7 or 73/5 or 101/47. However, every once in awhile we'll get a fraction that works out as a whole number, such as 5/5 (which equals 1) or 20/4 (which turns out as 5) or 1100/55 (which works

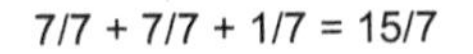

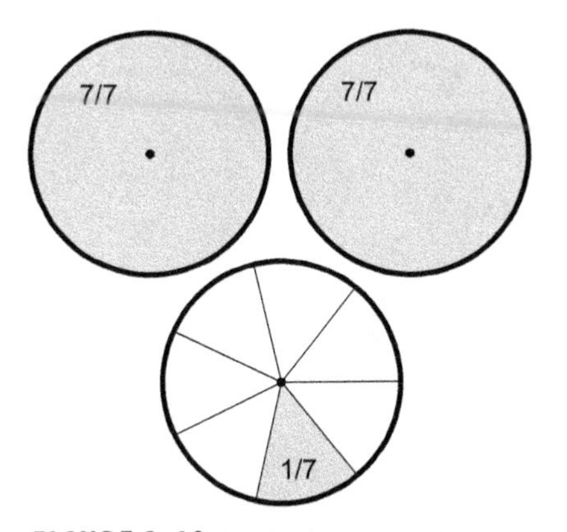

FIGURE 2-10 • We can portray a fraction larger than 1 as multiple objects plus part of another object. In this case we have two whole pies plus one-seventh of a third pie, giving us a total of two and one-seventh (or fifteen-sevenths) pies.

out as 20). Once we've "built up" the set of all possible fractions greater than or equal to 0, we can create a "mirror image" of the whole set by imagining a negative sign in front of every single value.

PROBLEM 2-12

Imagine that we have 12 guests at our dessert table. Of these people, 11 want pie, but they express (in no uncertain terms) that they're not only hungry, but ravenous. They each demand 1/4 of a pie! How many pies will we need, and how should we cut them up?

SOLUTION

We can take three pies and cut them each into fourths. When we do that, we'll use up two of the pies (4/4 plus 4/4), and 3/4 of the third pie, as shown in Fig. 2-11. The remaining 1/4 of the third pie can go into the refrigerator, and we can eat it at a later time (such as two seconds after the last dessert guest leaves). Alternatively, we can give that last 1/4 of a pie to the dessert guest who wasn't hungry for it. We can put the piece of pie in a "doggie bag" for that guest to take home, in case she (or her dog) wants to eat it later that evening.

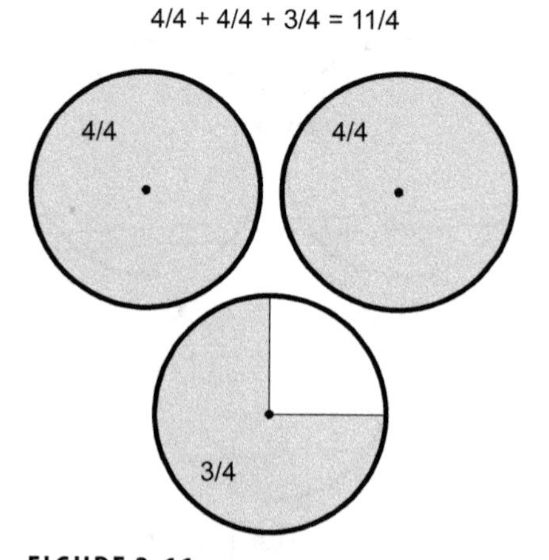

FIGURE 2-11 • Illustration for Problem 2-12 and its solution.

QUIZ

Refer to the text in this chapter if necessary. A good score is eight correct. Answers are in the back of the book.

1. Which of the following statements concerning even whole numbers doesn't always hold true?

A. We can split an even whole number into a product of primes.
B. If we cut an even whole number in half, we get a whole number.
C. If we cut an even number in half, we get an odd whole number.
D. If we add 1 to an odd whole number, we get an even whole number.

2. Which of the following numbers is not prime?

A. 2
B. 7
C. 37
D. 39

3. If we cut an odd number in half, we get

A. an even number.
B. a nonwhole fractional quantity.
C. an integer.
D. another odd number.

4. If we cut a prime number in half, we'll get

A. a whole number in one case, and nonwhole fractions in all the other cases.
B. an even number in one case, and odd numbers in all the other cases.
C. another prime number in every case.
D. a nonwhole fractional quantity in every case.

5. We can prove that two sets of numbers are the same size by

A. subtracting one from the other to get zero.
B. adding them to each other to get a new set that's twice as large as either of the original sets.
C. finding a one-to-one correspondence between them.
D. dividing one by the other to get 1.

6. Which of the following fractions constitutes a whole number?

A. 71/47
B. 123/41
C. 69/89
D. 20/40

7. **We call a counting number prime if**
   A. we can whole-number factor it into itself and 1, but in no other way.
   B. it's odd, and it's also an even multiple of some other odd number.
   C. it's even, and it's also an odd multiple of some other even number.
   D. Any of the above

8. **Which of the following numbers is prime? Use the test for prime factors described earlier in this chapter.**
   A. 3723
   B. 3717
   C. 3703
   D. 3697

9. **Suppose that we have three identical pies and 18 people, all of whom tell us that they're hungry for pie. We want to give each person an equal amount, and we also want to make sure that we give each person the maximum possible amount given the quantity of pie that we have on hand. We should cut each of the three pies up into**
   A. thirds.
   B. fourths.
   C. fifths.
   D. sixths.

10. **Imagine that, after we've cut up the pies and served them to our 18 dessert guests, two of those guests decide that they don't want their portion. They give their pieces of pie back to us. We're left with**
    A. 2/3 of a pie.
    B. 1/2 of a pie.
    C. 1/3 of a pie.
    D. 1/4 of a pie.

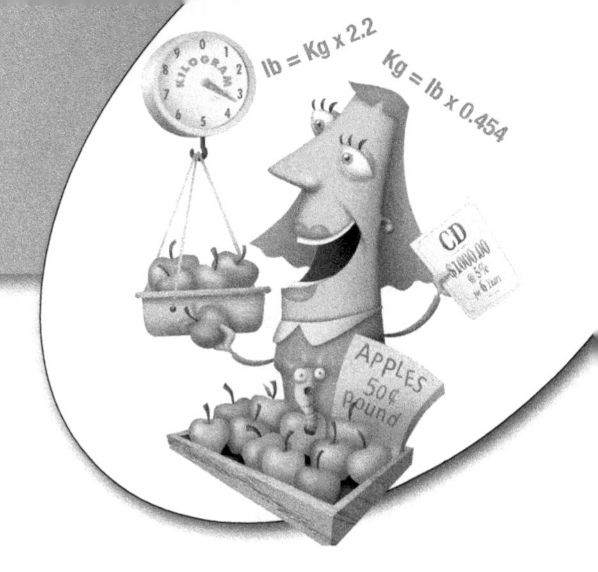

chapter 3

Decimals

Now that you know about fractions, let's look at another common way to symbolize numbers. Since the middle of the twentieth century, calculators and computers have taken over the mundane work of arithmetic. These machines work with *decimals*.

CHAPTER OBJECTIVES

In this chapter, you will

- Define and express quantities using powers of 10.
- Portray numbers geometrically along straight lines.
- Learn how to work with terminating decimal expressions.
- See how some fractions give rise to endless repeating decimals.
- Convert decimals to fractions and vice-versa.
- Discover endless decimals that lack patterns and can't be expressed as fractions.

Powers of 10

We can define a *positive-integer power* as a quantity multiplied by itself a certain number of times. If we divide a nonzero quantity by itself once, we say that we "raise" it to the *zeroth power*, and we always get 1 as the result. A *negative-integer power* of a certain number equals 1 divided by the number raised to the power with the minus sign removed.

We denote powers using superscripts called *exponents*, and we base decimal notation on integer powers of 10. In fact, some mathematicians call the number 10 the *common exponential base*. We can also call it the *base* or *radix* in the decimal numbering system.

Figure 3-1 shows a number line denoting several different powers of 10, ranging from 10^5 (the largest number here) down to 10^{-5} (the smallest). We call the relative difference between any two adjacent powers of 10 an *order of magnitude*. For example, 10^3 exceeds 10^2 by one order of magnitude, and 10^{-2} exceeds 10^{-5} by three orders of magnitude. At the extremes, 10^5 exceeds 10^{-5} by 10 orders of magnitude.

TIP ***The number line in Fig. 3-1 differs from "normal" number lines. All the numbers shown here have positive values. As you go upward on the line, the numerical value increases faster and faster, so you "race off toward infinity" more rapidly than you would do on a conventional number line. As you go downward, the value decreases at a slower and slower rate, "closing in on 0" but never getting all the way there.***

Still Struggling

If negative powers confuse you, consider the following examples. In each case, when you see 10 to the power of "negative *n*," it equals 1 divided by 10 to the power of "positive *n*," where *n* can stand for any positive whole number 1, 2, 3, 4, 5, and so on.

$$10^{-1} = 1/10^1 = 1/10$$

$$10^{-2} = 1/10^2 = 1/100$$

$$10^{-3} = 1/10^3 = 1/1000$$

This same rule holds for numbers other than 10. For example:

$$2^{-1} = 1/2^1 = 1/2$$

$$2^{-2} = 1/2^2 = 1/4$$

$$2^{-3} = 1/2^3 = 1/8$$

If you expand on the idea of Fig. 3-1, you can "build" positive numbers by taking single-digit multiples of powers of 10, and adding them up. Every positive number in this form has its negative "twin." To show the negative numbers, you can make up a separate number line for them. To account for 0, you can

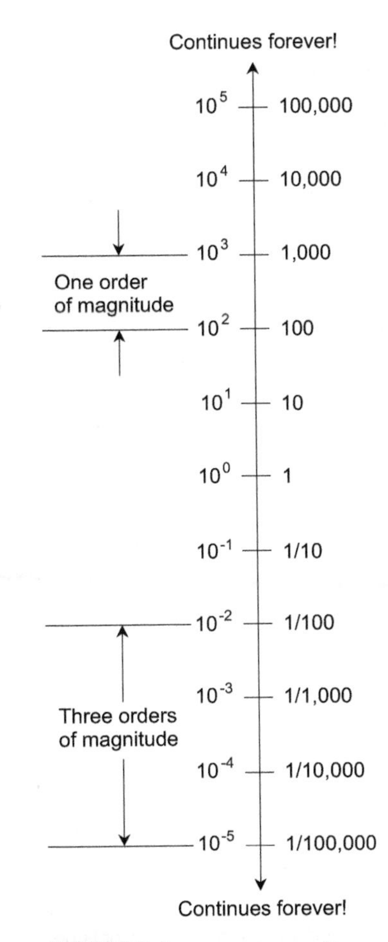

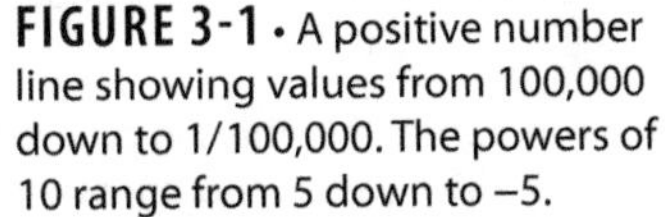

FIGURE 3-1 • A positive number line showing values from 100,000 down to 1/100,000. The powers of 10 range from 5 down to −5.

give it a special point all its own that doesn't lie on either line. Figure 3-2 shows numbers using this system.

The *decimal point* provides the foundation on which we can build up any decimal numeral we want. It looks like an ordinary period. We write the digits for positive powers of 10 to the left of the decimal point. We write the digits for negative powers of 10 to the right of the decimal point.

Let's break down the decimal numeral 362.7735. Starting at the decimal point and working toward the left, we get

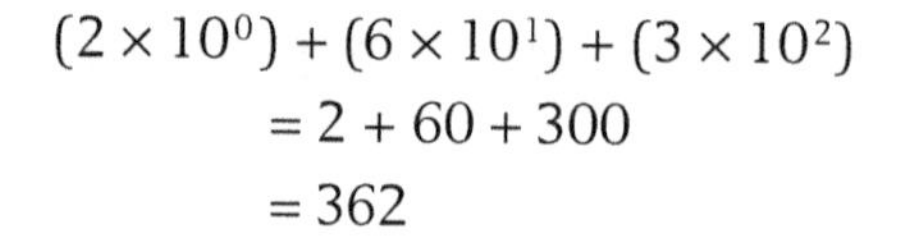

$$\begin{aligned}(2 \times 10^0) + (6 \times 10^1) + (3 \times 10^2) \\ = 2 + 60 + 300 \\ = 362\end{aligned}$$

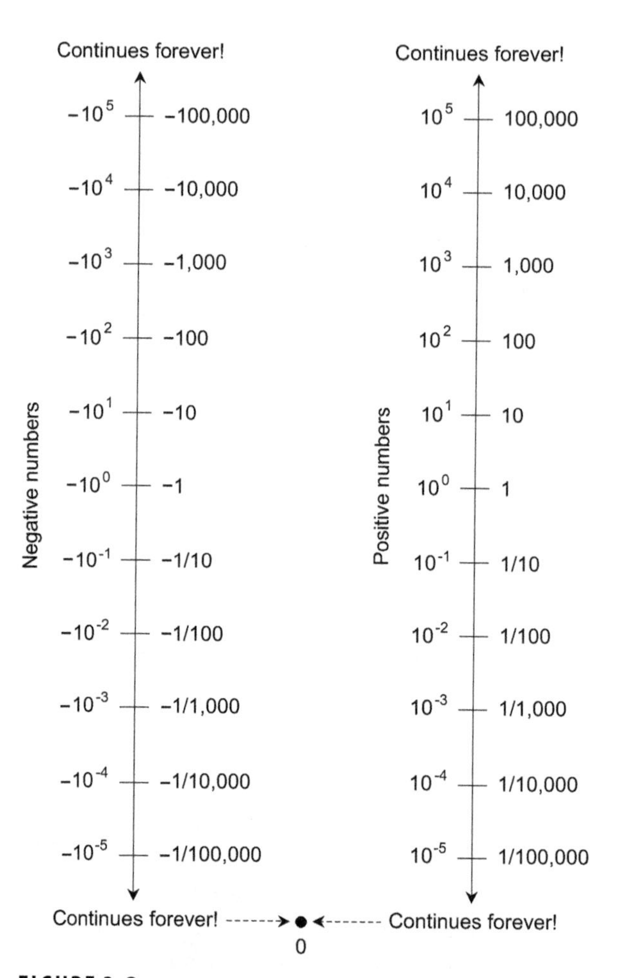

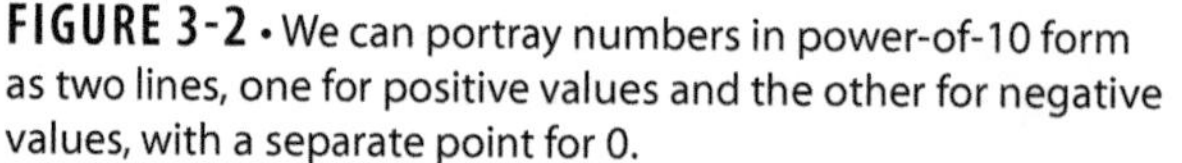

FIGURE 3-2 • We can portray numbers in power-of-10 form as two lines, one for positive values and the other for negative values, with a separate point for 0.

Starting at the decimal point and working toward the right, we get

$$(7 \times 10^{-1}) + (7 \times 10^{-2}) + (3 \times 10^{-3}) + (5 \times 10^{-4})$$
$$= 7/10 + 7/100 + 3/1000 + 5/10{,}000$$
$$= 0.7735$$

When we add the whole number to the decimal fraction, we end up with

$$362 + 0.7735 = 362.7735$$

You'll sometimes hear "science geeks" talk about orders of magnitude involving both positive and negative numbers. They mean to say that the *absolute value* of one quantity is some power of 10 times bigger or smaller than the *absolute value* of the other quantity. Consider these examples:

- 450 is one order of magnitude larger than 45
- 0.56 is two orders of magnitude smaller, in absolute terms, than −56
- − 0.565 is three orders of magnitude smaller, in absolute terms, than 565
- − 88,888 is four orders of magnitude larger, in absolute terms, than − 8.8888

If one or both quantities are negative, you should always include the phrase "in absolute terms" so that nobody gets confused about the meanings of "smaller" or "larger." When you want to portray a decimal number greater than or equal to 0 but smaller than 1, you should write a single numeral 0 to the left of the decimal point. If the number is greater than −1 but smaller than 0, you should write the minus sign first, then a single 0, and then the decimal point.

PROBLEM 3-1

Draw a number line in power-of-10 style that shows numbers from 10 to 100,000. How many orders of magnitude does this line show?

SOLUTION

Figure 3-3 portrays a number line that covers the range of positive values from 10 up to 100,000. To calculate the span in orders of magnitude, subtract the smallest power of 10 from the biggest power 10. Here, you have

$$100{,}000 = 10^5$$

and

$$10 = 10^1$$

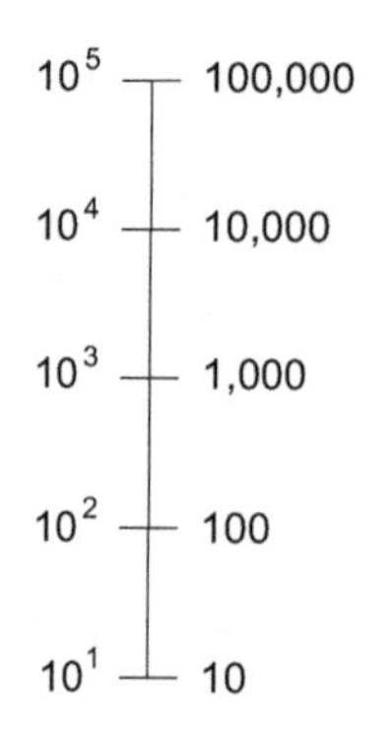

FIGURE 3-3 • Illustration for Problem 3-1 and its solution.

You get 5 – 1, or 4, orders of magnitude in this span. (You can also see the number of orders of magnitude on this line by counting the number of intervals between "hash marks"—one, two, three, four!)

PROBLEM 3-2

Draw a number line in power-of-10 style that shows numbers from 30 to 300,000. How many orders of magnitude does it show?

SOLUTION

Figure 3-4 portrays a number line that covers the range of positive values from 30 up to 300,000. In this case, you can divide the larger number

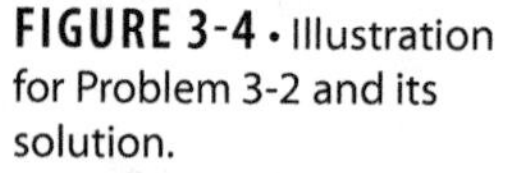

FIGURE 3-4 • Illustration for Problem 3-2 and its solution.

by the smaller, and then count the number of ciphers in the quotient. You get

$$300{,}000/30 = 10{,}000$$

These two numbers differ by four orders of magnitude, the same extent as the two numbers differ in Problem 3-1. (You can also count the number of intervals between hash marks, just as you might have done to solve Problem 3-1.)

Terminating Decimals

An everyday decimal expression always has a finite number of digits to the right of the decimal point. One common example is the notation for American dollars and cents, where you'll see two digits "beyond" the decimal point. Any further digits, if you want to write them, all turn out as ciphers. We call this sort of numeral a *terminating decimal.*

In a terminating decimal, the digits to the right of the decimal point always represent a fraction having a *denominator* (lower or second part) that equals a power of 10. If you see one digit, it represents 10ths; if you see two digits, they represent 100ths; if you find three digits, they represent 1000ths; and so on, as far as you want to go. For example:

- 0.7 represents 7/10
- 0.72 represents 72/100
- 0.729 represents 729/1000
- 0.7294 represents 7294/10,000
- 0.72941 represents 72,941/100,000

You can also write the above examples as

- 0.7 represents 7×10^{-1}
- 0.72 represents 72×10^{-2}
- 0.729 represents 729×10^{-3}
- 0.7294 represents 7294×10^{-4}
- 0.72941 represents $72{,}941 \times 10^{-5}$

TIP *If you want to reduce a quantity by exactly* n *orders of magnitude, move the decimal point* n *places to the left. If you want to increase a quantity by exactly* n

orders of magnitude, move the decimal point* n *places to the right. If necessary, add extra zeros on the left and right to serve as "placeholders."

Still Struggling

The foregoing examples *do not* represent changes in the order of magnitude. Instead, they represent numbers whose values lie close to each other. The values approach the last number in either list: 0.72941. If that's as far as you want to go, then adding any more digits to the right will only clutter up the page with unnecessary ciphers.

The following numerals all represent exactly the same number to a mathematician interested only in pure theory:

0.72941

0.729410

0.7294100

0.72941000

0.729410000

0.7294100000

↓

and so on, forever

Physicists or engineers see the above numbers from a different point of view than the pure mathematician does. To a "nuts-and-bolts" person, the extra ciphers represent increasing precision. They constitute *significant figures*.

In any decimal expression, the number of digits to the left of the decimal point is always finite. If you "chop off" all the digits to the right of the decimal point, the digits to the left of the point represent an integer. You might add ciphers to the left-hand end of the digit string without changing the value, but you'll rarely have any reason to do that. For example, you won't often see a numeral such as

00,000,004,580,103.7864892022

Instead, you'd see it written as

4,580,103.7864892022

TIP *You should never expect to see commas to the right of a decimal point, no matter how far out the digits go; and when you write an expression that needs a lot of digits to the right of the decimal point, you should never insert any commas there. Commas belong only on the left-hand side of a decimal point.*

Still Struggling

Numbers such as 4,580,103.7864892022 can be hard to read, especially when you see them on a calculator display that doesn't include commas. You can insert spaces on either side of the point, so the number reads as

4,580,103 . 7864892022

You can also insert a space after every third digit to the right of the decimal point. The above numeral will then appear as

4,580,103 . 786 489 202 2

Use caution when you insert spaces into a numeral! In this example, the lonely digit 2 at the end of the "digit string" might confuse some people. A few folks might not even see it at all.

PROBLEM 3-3

Break down 36.573 into the sum of a single integer and a single fraction.

SOLUTION

Let's start by paying attention only to the digits on the left-hand side of the decimal point. They form the integer 36. On the right-hand side of the point, we see 573, a string of three digits. Therefore, we should denote the fraction's denominator as a 1 with three ciphers after it. When we put the digits following the decimal point into the *numerator* (upper or first part) of the fraction, we get

573 / 1000

The entire quantity "breaks down" into the sum

36 + (573 / 1000)

We add the parentheses around the fraction for clarity.

TIP *If you're a perfectionist who remembers everything from your algebra textbooks, you'll recall the fact that the above parentheses aren't technically necessary. However, they don't do any harm, and they can help you to avoid confusion.*

PROBLEM 3-4

Break down 4,580,103 . 7864892022 into the sum of a single integer and a single fraction.

SOLUTION

As we did in the solution to Problem 3-3, let's start by examining only the digits to the left of the decimal point. That digit sequence portrays the integer

$$4{,}580{,}103$$

Now let's look to the right of the point. We find 10 digits here, so we should denote the fraction's denominator as a numeral 1 with 10 ciphers after it. When we put the digits following the decimal point into the numerator, we get the complete fraction

$$7{,}864{,}892{,}022 / 10{,}000{,}000{,}000$$

The entire quantity 4,580,103 . 7864892022 equals the sum of the foregoing integer and the foregoing fraction, so we have the "breakdown sum"

$$4{,}580{,}103 + (7{,}864{,}892{,}022 / 10{,}000{,}000{,}000)$$

As before, we add parentheses for clarity (so nobody gets the mistaken idea that they should add 4,580,103 to 7,864,892,022, and *then* divide by 10,000,000,000).

Endless Repeating Decimals

Whenever you write out a "real-world" decimal expression, you deal with a *terminating decimal*, meaning that after a certain place in the digit string, you have only ciphers (zeros), so you have no reason to keep on going. But in certain theoretical situations, the digits to the right of a decimal point can continue without end, and you'll *never* reach a spot where every digit to the right is a cipher. In a situation of that sort, you have an *endless repeating decimal*, also called a *nonterminating repeating decimal*.

TIP ***Your calculator can give you a good idea of what an endless repeating decimal looks like. The calculator program in a personal computer works great for this purpose because it displays a lot of digits. (Of course, you can never envision all of the digits in an endless repeating decimal. Nobody can "see all the way to infinity"!)***

If you want to get a good idea of the distinction between a terminating decimal and an endless repeating decimal, you can use the "$1/x$" key on your calculator and start with 2 for x. Then try it with 3, 4, 5, and so on, watching the results:

$$1/2 = 0.5$$

$$1/3 = 0.333333333333 \cdots$$

$$1/4 = 0.25$$

$$1/5 = 0.2$$

$$1/6 = 0.166666666666 \cdots$$

$$1/7 = 0.142857142857142857 \cdots$$

$$1/8 = 0.125$$

$$1/9 = 0.111111111111 \cdots$$

The fractions 1/2, 1/4, 1/5, and 1/8 all work out as terminating decimals. The fractions 1/3, 1/6, 1/7, and 1/9 divide out as endless repeating decimals. The presence of an *ellipsis* (three small dots) indicates that the pattern continues forever. Note the uniqueness of 1/7, which goes through a repeating cycle of the six digits 142857.

TIP ***Once in awhile you'll see a situation where both the numerator and denominator of a fraction are integers, but you can't find any pattern in the decimal expression because the repeating sequence has too many digits. Have faith! A pattern exists, whether you can see it straightaway or not.***

Still Struggling

When you divide an integer by another integer, and if the two integers are large enough, your calculator display might not show enough digits to let you see the pattern of repetition. Take a calculator that can show only 10 digits, and divide out the fraction

138,297,004,792,676 / 999,999,999,999,999

Suppose that you show your calculator display to a friend, tell her it's the quotient of two integers that you entered, and then ask her to tell you what fraction you put in. Even if she has a Ph.D. in mathematics, she'll never figure it out. The pattern here exceeds the capacity of the electronic display. If your calculator puts a zero to the left of the decimal point and rounds off the last digit, you and your friend will see something like

0.138297005

which obviously doesn't tell the whole story.

PROBLEM 3-5

Write the following ratios as decimal expressions.

44/16

−81/27

51/13

−45/800

SOLUTION

You can convert each of these ratios to decimals by inputting the numerator into a calculator, and then dividing by the denominator. When you go through those maneuvers, you get the following results.

44/16 = 2.75

−81/27 = −3

51/13 = 3.923076923076923076 ⋯

−45/800 = −0.05625

In the case of 51/13, you'll need a calculator that can display a lot of digits if you want to make sure that you know the repeating pattern of digits to the right of the decimal point. (It's 923076.) If you don't have such a calculator, you can perform old-fashioned, manual long division to discover the pattern—but it'll take you quite awhile!

PROBLEM 3-6

Convert the fraction 1/17 to another fraction whose denominator comprises a string of 9s.

SOLUTION

You can solve this problem in two steps. First, use a calculator or long division to determine the decimal equivalent of 1/17. You'll get

0.05882352941176470588235294117647 ···

The repeating sequence of digits is 0588235294117647. The initial cipher plays an important role here! Next, count up the number of digits in this sequence, including the cipher at the beginning. You'll find 16 digits in the repeating string. Construct a fraction with the repeating sequence in the numerator and a string of 16 digits, all 9s, in the denominator, inserting commas to make the large numbers easy to read. You'll get

0,588,235,294,117,647 / 9,999,999,999,999,999

You can now remove the initial cipher. (You needed it to ensure that you put the correct number of 9s in the denominator; but now it's served that purpose, so you can get rid of it.) You have the final result

588,235,294,117,647 / 9,999,999,999,999,999

If you want, you can divide out this fraction using a calculator with a large display, such as the one in a computer. You should get the same result as you got when you divided 1 by 17.

TIP *As an alternative way to check the solution to Problem 3-6, use your computer's "many-digit" calculator to divide the denominator by the numerator (in other words, do the operation "upside-down"). If you hit all the keys right, you'll get exactly 17.*

PROBLEM 3-7

Imagine a decimal expression that has an endlessly repeating triplet of digits. We can write it down in this form:

$$0\,.\,\#\#\#\ \#\#\#\ \#\#\#\ \cdots$$

where ### represents the sequence of three digits that repeats. We insert spaces on either side of the decimal point, and after each triplet of pound signs, to make the expression clear. Our mission: Show that this decimal numeral represents the fraction

$$\#\#\#\,/\,999$$

SOLUTION

Let's call the "mystery fraction" m. We want to find a fractional expression for m. This process takes several steps. Follow along closely to understand how it works. We've been told that

$$m = 0\,.\,\#\#\#\ \#\#\#\ \#\#\#\ \cdots$$

We can break this expression down into a sum of two decimals, one terminating and the other endless, as follows:

$$m = (0\,.\,\#\#\#) + (0\,.\,000\ \#\#\#\ \#\#\#\ \#\#\#\ \cdots)$$

Note that the first addend here is ### / 1,000. The second addend happens to be the original mystery number, m, divided by 1,000. Therefore,

$$m = (\#\#\#\,/\,1{,}000) + (m\,/\,1{,}000)$$

We can add the two fractions on the right-hand side of the equals sign to get

$$m = (\#\#\# + m)\,/\,1{,}000$$

Let's multiply each side of the equation by 1,000 and then manipulate the right-hand side to obtain

$$\begin{aligned} 1{,}000\,m &= 1{,}000\,(\#\#\# + m)\,/\,1{,}000 \\ &= (\#\#\# + m)\,(1{,}000\,/\,1{,}000) \\ &= \#\#\# + m \end{aligned}$$

We can simplify to get

$$1{,}000\,m = \#\#\# + m$$

When we subtract m from the expressions on both sides of the equals sign, we have

$$1{,}000m - m = \#\#\# + m - m$$

which we can simplify to

$$999\,m = \#\#\#$$

Finally, let's divide each side by 999. When we do that, we end up with

$$m = \#\#\# / 999$$

Mission accomplished!

Still Struggling

When you see a fraction whose denominator comprises a finite string of 9s and the numerator is less than the denominator, you know that the fraction converts to an endless repeating decimal expression with 0 to the left of the point, and a repeating sequence of digits to the right of the point. But there's a catch! The repeating sequence to the right of the point must contain the same number of individual digits as the denominator has. If the numerator has fewer digits than the denominator contains, you must add ciphers "in front of" (on the left end of) the numerator until it has the same number of individual digits as the denominator contains.

PROBLEM 3-8

Convert the fraction 57 / 99 to a repeating decimal.

SOLUTION

In this case, the denominator has the same number of digits as the numerator does, so we can immediately tell that the repeating decimal is

$$57/99 = 0.575757 \cdots$$

We can verify this fact with a calculator.

PROBLEM 3-9

Convert the fraction 57/9999 to a repeating decimal.

SOLUTION

This fraction has two digits in the numerator, but four in the denominator. Therefore, we should convert it to 0057/9999 before we put down the repeating sequence of digits. Then we can confidently write

$$57/9999 = 0.005700570057\cdots$$

As in the previous problem, we can verify this fact with a calculator, preferably one that has a display with a lot of digits!

Terminating Decimal to Fraction

When you see a ratio of integers, you can convert it to decimal form if you have a calculator that can display enough digits. But that's the catch! Even a good calculator can fall short in this respect. If you have a calculator with a 10-digit display and you divide 1 by 7, you won't even see two full repetitions of the pattern. If you didn't know better from having seen the decimal expansion of 1/7 earlier in this chapter, you might never deduce it from a 10-digit calculator alone. If you have a good computer calculator program, you're better off. But even the best calculators can "choke" if you give them a "bad" enough ratio. Try 51/29, for example!

Fortunately, you won't have to perform ratio-to-decimal conversions very often. When you must do it, the calculator program in any good personal computer will usually work. In the extreme, you can always resort to old-fashioned, manual longhand division (which you'll get to review in Chap. 5). You can also write, or find, a computer program to "grind out" thousands of digits and look for patterns.

When you see a terminating decimal expression and you want to convert it to a ratio of one integer to another, you can do it in steps. Here's an example. Imagine that you see the following decimal numeral and you want to put it into ratio form as a quotient of two integers:

$$3588.7601811$$

Move the point to the right until it's at the end of the string of digits, leaving nothing beyond. Then delete the point. You'll get the whole number

35,887,601,811

Now make this number the numerator of a fraction. Count the number of places you moved to the right to get the point to the end of the string of digits. (In this case, it's seven places.) Then in the denominator of the fraction, write down a 1 followed by that number of ciphers. The result:

35,887,601,811 / 10,000,000

TIP *You can apply the foregoing decimal-to-fraction conversion method to any terminating decimal expression that you'll ever see.*

PROBLEM 3-10

Write each of the following decimal expressions as the ratio of an integer to a counting number.

4.7

−8.354

0.000022

−0.0792

SOLUTION

First, let's convert 4.7 to a fraction. When we move the decimal point to the end of the string of digits and then delete it, we get 47, which will serve as the numerator of our fraction. Next, let's recall that we moved the decimal point one place to the right to get rid of it. Our denominator, therefore, comprises the digit 1 followed by a single cipher, so it's 10. We've just figured out that

$$4.7 = 47/10$$

Next, let's convert −8.354 to a fraction. We must move the decimal point three places to the right to get rid of it, leaving us with −8354. We make this number the numerator of our fraction. Our denominator contains the digit 1 followed by three ciphers, so it's 1000, and we know that

$$-8.354 = -8354/1000$$

Now we come to 0.000022. To eliminate the decimal point, we must shift it six places to the right. Then we get 0000022, which equals 22. We make it the numerator of our fraction. Our denominator starts with 1, and then we must add six ciphers to get 1,000,000. We've just figured out that

$$0.000022 = 22/1{,}000{,}000$$

Finally, let's deal with −0.0792. When we move the decimal point four places to the right, we get −00792, which is the same thing as −792. That's our numerator. As for the denominator, we write down 1 and then four ciphers, obtaining 10,000. We can conclude that

$$-0.0792 = -792/10{,}000$$

TIP ***In all of the preceding conversions, you'll see positive denominators in the final answers. By convention, that's how you should always portray a fraction. If you ever encounter a fraction that has a negative denominator, you should remove the minus sign from the denominator, and then make the numerator equal to the negative of its original value (negative to positive, or positive to negative).***

Endless Repeating Decimal to Fraction

When you encounter a decimal expression containing a sequence of digits that repeats without end and you want to convert it to a fraction, first split the integer part from the decimal part. Call the integer part a. Then write down the part of the expression to the right of the decimal point, with the point on the extreme left. You'll get an expression of the form

$$.\,b_1 b_2 b_3 \cdots b_n b_1 b_2 b_3 \cdots b_n b_1 b_2 b_3 \cdots b_n$$

where $b_1\ b_2\ b_3 \cdots b_n$ represents the repeating sequence, which contains n digits. (Each b with a subscript represents a single digit.) The fractional part of the expression is

$$b_1 b_2 b_3 \cdots b_n/999 \cdots 999$$

where the denominator has n digits, all 9s. Now you can put back the integer part, getting the number in the form

$$a\text{-}b_1 b_2 b_3 \cdots b_n/999 \cdots 999$$

Here, the dash after the a serves only to separate the integer part of the expression from the fractional part. It's *not* a minus sign! If this notation confuses you, put a plus sign in place of the dash. Then you'll get the equivalent expression

$$a + b_1 b_2 b_3 \cdots b_n / 999 \cdots 999$$

If you want to convert an endless, repeating decimal expression to a "pure fraction" without any integer component, you must convert the integer a to a fraction with a denominator comprising all 9s. Multiply a by 999 ··· 999, put the result into the numerator of a new fraction, and then put 999 ··· 999 in the denominator, getting

$$999 \cdots 999 \times a / 999 \cdots 999$$

Add this new fraction to the fraction you got when you converted the decimal part of the original expression. You'll get something of the form

$$(999 \cdots 999 \times a / 999 \cdots 999) + (b_1 b_2 b_3 \cdots b_n / 999 \cdots 999)$$

Finally, add these two "monsters" to get the pure fraction

$$[(999 \cdots 999 \times a) + (b_1 b_2 b_3 \cdots b_n)] / 999 \cdots 999$$

PROBLEM 3-11

Convert the following endless decimal to a pure fraction:

$$\mathbf{23.860486048604 \cdots}$$

SOLUTION

The decimal portion is a sequence of the digits 8, 6, 0, and 4 that endlessly repeats. We can tell right away that this is 8604 / 9999. The whole-number portion, 23, can be multiplied by 9999, and the result put into the numerator of a fraction. The denominator should then be 9999, so we get

$$(23 \times 9999) / 9999$$

$$= 229{,}977 / 9999$$

That's just the whole number 23 expanded into 9,999ths. Now we add the decimal part back in, so the entire number becomes

$$229{,}977 / 9999 + 8604 / 9999$$

$$= (229{,}977 + 8604) / 9999$$

$$= 238{,}581 / 9999$$

If you use a calculator that can display a lot of digits to divide out this fraction, you should get the original expression: 23 followed by a decimal point, and then the sequence of digits 8, 6, 0, and 4 repeating.

PROBLEM 3-12

Convert the following endless decimal to a pure fraction:

$$0.248624862486\cdots$$

SOLUTION

Here, we have the repeating sequence of digits 2486. We make it the numerator of a fraction, and then put an equal number of 9s (in this case four) in the denominator. When we do that, we discover that

$$0.248624862486\cdots = 2486/9999$$

You can check it out with a calculator whose display can show a lot of digits. Input 2, 4, 8, and 6 in that order, then hit the "divide by" button, then enter 9, 9, 9, and 9, and finally hit the "equals" button.

PROBLEM 3-13

Convert the following endless decimal to a pure fraction:

$$-2.892892892\cdots$$

SOLUTION

Whenever you encounter a negative endless decimal quantity and you want to convert it to a fraction, you must take care with the signs! The minus sign applies to the entire quantity, both the integer part and the fractional part. If you get sloppy, you might add in the fraction instead of taking it away. Before you start working out the conversion of a negative endless decimal expression, delete the minus sign and treat all values as if they were positive. Keep treating all values as positive throughout the process. Then, when you know that you've done all the arithmetic right, put the minus sign back in *as the very last step*. Following that "program" in this case, you should start with the quantity

$$2.892892892\cdots$$

Write down the part of the expression to the right of the decimal point, getting

$$.892892892 \cdots$$

The repeating sequence is 892, so the fractional part equals 892 / 999. Now you can put in the integer part without the minus sign, getting

$$2 + 892/999$$

Multiply 2 by 999, getting 1998, and make it the numerator of a new fraction whose denominator equals 999. Now you have two fractions that you can add to get

$$892/999 + 1998/999 = (892 + 1998)/999$$
$$= 2890/999$$

Finally, you come to the last step, which you had better not forget! Put the minus sign back in. You'll end up with the fact that

$$-2.892892892 \cdots = -2890/999$$

TIP ***Always check the results of maneuvers such as those in the preceding problems by dividing out your final fraction using a calculator with a long display. Make certain that the resulting quotient equals the original endless decimal expression.***

Endless Nonrepeating Decimals

You might wonder if any decimals exist that go on without end, but that *never* produce a repeating sequence. Indeed they do exist, and you don't have to search very hard to find examples.

Consider the circumference of a perfect circle divided by its diameter. This value never varies, no matter how large or small the circle gets, as long as the circle and its interior lies entirely on a flat surface. In ancient times, people knew that any circle's circumference slightly exceeds three times its diameter. People tried to define the relationship as an exact fraction, but no one could ever quite get it right. For a long time, people thought that 22/7 represented the exact ratio, and even today, it's good enough for rough estimates. Eventually, however, some astute mathematicians proved that a circle's circumference divided by its diameter, so easily definable in terms of geometry, doesn't work out precisely as the ratio of *any* two whole numbers.

If you've taken a basic geometry course, you know that the circumference of a circle divided by its diameter is symbolized by the lowercase Greek letter *pi* (π). Many calculators have a key you can punch to get π straightaway. You'll get 3.14159 followed by an apparently random string of digits that runs all the way to the end of the display.

Still Struggling

Even if you spend the rest of your life trying to find a repeating sequence of digits in the decimal expansion of π, you'll fail. Various other quantities also act this way when you try to write out their decimal expansions. Mathematicians call them *irrational numbers*, meaning that you can't express them as *ratios* of integers.

QUIZ

Refer to the text in this chapter if necessary. A good score is eight correct. Answers are in the back of the book.

1. **Which of the following quantities differs from 2 by an order of magnitude?**
   A. 20
   B. 2/10
   C. 0.2
   D. All of the above

2. **Consider the decimal numeral 23.456. If we move the decimal point two places to the right, we get a new quantity that's**
   A. twice as great as the original quantity.
   B. 10 times as great as the original quantity.
   C. 100 times as great as the original quantity.
   D. no different from the original quantity.

3. **Consider the decimal numeral 78,003.22. If we move the decimal point four places to the left, we get a new quantity that's**
   A. 1/4 as great as the original quantity.
   B. four orders of magnitude smaller than the original quantity.
   C. four orders of magnitude larger than the original quantity.
   D. no different from the original quantity.

4. **We can write the fraction 3456/10 in decimal form as**
   A. 345.6
   B. 34.56
   C. 3.456
   D. 0.3456

5. **We can write the fraction 3456/1000 in decimal form as**
   A. 345.6
   B. 34.56
   C. 3.456
   D. 0.3456

6. **In decimal form, we can write the fraction 2371/9999 as**
   A. 0.2371
   B. 0.237123712371 ···
   C. 2.371371371 ···
   D. 2.371

7. In decimal form, we can write the fraction 23 / 9999 as

A. 0.232323 ···
B. 0.023023023 ···
C. 0.002300230023 ···
D. 0.000230002300023 ···

8. The endless repeating decimal 0.888 ··· equals the fraction

A. 8 / 9
B. 88 / 99
C. 888 / 999
D. All of the above

9. Suppose that we encounter the following endless decimal expression, where the number of ciphers between the 1s keeps doubling as we move out toward the right:

0.0100100001000000001 ···

This quantity equals the fraction

A. 1 / 99
B. 1 / 9999
C. 1 / 99,999,999
D. None of the above

10. From the viewpoint of a mathematician interested in pure theory, what's the difference between 24.56 and 24.560?

A. The cipher in the second expression represents an additional significant figure.
B. The first expression is more precise than the second one.
C. The second expression is more precise than the first one.
D. A mathematician interested in pure theory sees no difference between the two expressions.

chapter 4

Proportions

A *ratio* tells us how two quantities compare in terms of size. A *proportion* expresses how a quantity changes in value when another, related quantity varies. Let's learn how ratios and proportions relate to fractions. Then we'll see how we can use ratios to define certain numbers.

CHAPTER OBJECTIVES

In this chapter, you will

- Learn how fractions relate to ratios.
- Contrast proportions with fractions and ratios.
- Compare direct and inverse proportionality.
- Convert fractions to percentages.
- See how percentages can help you quantify gains or losses.
- Distinguish between rational and irrational numbers.
- Define the collection of real numbers.

Ratios

You can express any ratio as a fraction, even though fractions and ratios technically differ from one another. If you say, "We must travel *eight-fifths* as far to get to the town of Happyville as we have to go to reach the town of Bluesdale," then you use 8/5 as a fraction. If you say, "The ratio of the distance to Happyville compared with the distance to Bluesdale equals *eight to five*," then you say the same thing in terms of a ratio.

You can write, "The ratio of the distance to Happyville compared with the distance to Bluesdale equals 8/5," and read "8/5" as "eight to five." If you want to clarify the fact that you want to talk about a ratio, you can use a colon instead of a slash to separate the 8 and the 5 and write, "The Happyville-to-Bluesdale distance ratio equals 8:5," reading "8:5" as "eight to five."

Ratios don't always have fixed numerators and denominators. Maybe you have to drive 80 miles to get to Happyville, but only 50 miles to get to Bluesdale. Or maybe you have to make a journey of 1208 kilometers to reach Happyville and 755 kilometers to reach Bluesdale. Either way, the distance ratio equals 8:5. In the first case, $80 = 8 \times 10$ and $50 = 5 \times 10$. In the second case, $1208 = 8 \times 151$ and $755 = 5 \times 151$. When you divide 80 by 50, or when you divide 1208 by 755, you get 1.6, which equals 8/5.

TIP ***In most situations, you'll want to express a ratio in terms of an integer (which can be either positive or negative) divided by a positive integer. Sometimes you can express a ratio in terms of a decimal quantity to 1 (for example 1.6:1), or as 1 to a decimal quantity (such as 1:1.6). However, you should never express a ratio as an integer-and-fraction quantity to 1. You will cause a lot of confusion if you write the foregoing ratio as 1-3/5:1, or even worse, 1-3/5/1.***

PROBLEM 4-1

A motor vehicle dealer tells you that a Model X sports car has a top speed that is "half again" as fast as the top speed of a Model Y truck. Figure out the ratio of the car's top speed compared to the truck's top speed in terms of positive integers. Then go the other way, and figure out the ratio of the truck's top speed to the car's top speed in terms of positive integers.

SOLUTION

The term "half again" means "one and a half times as much." The ratio of the car's top speed to the truck's top speed, therefore, equals 1-1/2 to 1, or 3/2

to 1, or 1.5:1. None of these expressions give you an integer divided by a positive integer! If you want to stick with that form, the best way to express the ratio is 3 to 2. You can write it as 3/2 or 3:2. When you want to express a ratio in the reverse sense, simply swap the numerator and the denominator. The ratio of the truck's top speed to the car's top speed equals 2 to 3, which you can write as 2/3 or 2:3.

Still Struggling

You might say that the ratio of the car's top speed to the truck's top speed is 6:4 or 9:6 or even something like 69:46, as long as the left-hand value equals 1.5 times the right-hand value. Maybe the car can go up to 120 miles an hour, and the truck can go up to 80 miles an hour, so you'd call the ratio 120:80. Conversely, you could say that the ratio of the truck's top speed to the car's top speed is 10:15 or 20:30 or 80:120. But exercises of this sort can get ridiculous. Normally, you'll want to express a ratio between an integer and a positive integer in the simplest possible way (technically called the *lowest form*), where the values are as small as possible (in this case 3:2 or 2:3).

Direct Proportions

We can use proportions to demonstrate the fact that two ratios, even though written differently, actually equal each other. We can also use proportions to say that a certain quantity increases when (and because) another one increases, or that a certain quantity goes down when (and because) another one goes up.

Let's compare a few ratios. Look carefully at the relative values of the first and second numbers. Can you see the pattern?

1:2

2:4

3:6

4:8

5:10

6:12

In each case, the denominator equals twice the numerator. We can write each of these ratios as the fractional value 1/2 because they all actually equal 1/2! Because one quantity changes, multiplication-wise, precisely along with the other one, we say that the right-hand quantity *varies in direct proportion to* the left-hand quantity. Some people say that the two quantities *are directly proportional* to each other. If we want to get specific, we can make claims such as the following:

- 1 is to 2, as 2 is to 4
- 1 is to 2, as 3 is to 6
- 1 is to 2, as 4 is to 8
- 1 is to 2, as 5 is to 10
- 1 is to 2, as 6 is to 12

We can juggle the numbers and get some more valid statements, including:

- 2 is to 4, as 3 is to 6
- 3 is to 6, as 4 is to 8
- 4 is to 8, as 5 is to 10
- 5 is to 10, as 6 is to 12

The above expressions all constitute *statements of proportionality*. If we allow the values to vary, always making sure that the right-hand value (call it R) equals twice the left-hand value (call it L), we can write

$$L:R = L/R = 1/2 = 1:2$$

In a sequence of ratios having a direct proportion, the quantities don't have to keep increasing. They can both decrease instead. For example, the following ratio sequence has direct proportionality:

40:20

38:19

36:18

34:17

32:16

30:15

In each ratio, the left-hand quantity equals twice the right-hand quantity.

Still Struggling

When mathematicians or scientists want to write that a certain quantity (say x) varies in direct proportion to another quantity (say y), they sometimes use a symbol that looks like a cut-off, sideways 8 between them. The proportionality expression

$$x \propto y$$

translates literally into the words "x is proportional to y," or more precisely, "x varies in direct proportion to y."

Inverse Proportions

Now let's examine a different situation. Once again, note the relative values of the first and second numbers. Can you find the pattern?

64:1

32:2

16:4

8:8

4:16

2:32

1:64

As we move down the list, the left-hand quantity keeps getting half as big as it was before, and the right-hand quantity keeps doubling. Obviously, we can't write all of these ratios as the same fraction! But we can see a pattern. One quantity goes up at the same rate, multiplication-wise, as the other one goes down. In a situation like this, we say that the right-hand quantity *varies in inverse proportion to* the left-hand quantity. Alternatively, we can say that the two quantities *are inversely proportional to* each other.

PROBLEM 4-2

Suppose that a truck travels west on a straight highway at a speed of 50 miles per hour (mph). A car going 60 mph passes the truck. Then a motorcycle passes

the car, at a speed that exceeds the car's speed in the same proportion as the car's speed exceeds the truck's speed. How fast is the motorcycle going?

SOLUTION

Let's calculate the proportional difference between the car's speed and the truck's speed. We divide 60 by 50 to get a ratio of 60:50, which we can reduce to 6:5 or 1.2:1. Now we know that the car moves 1.2 times as fast as the truck does, so the motorcycle must move 1.2 times as fast as the car does. That's 1.2 × 60 mph, or 72 mph.

PROBLEM 4-3

State the proportionality relation among the truck, car, and motorcycle speeds in the solution to Problem 4-2, in the form of an "is-to" statement. Then reverse that statement to get another "is-to" statement in the opposite sense. What ratios correspond to these statements?

SOLUTION

We can say "72 mph is to 60 mph, as 60 mph is to 50 mph." We can turn it around and say "50 mph is to 60 mph, as 60 mph is to 72 mph." In the first case, we talk about ratios of 6:5 or 1.2:1. In the second case, we talk about ratios of 5:6 or 1:1.2.

Percentages

The *percentage* method allows you to express fractional increases or decreases in a modified decimal format. Percentages came into existence before decimals did. Some people find decimals easier to understand than percentages, but mathematicians relied on percentages for many years, especially in economics and statistics. You'll still see them a lot.

The idea of percentage arose because fractional expressions can grow clumsy, and comparing their values can get difficult. If someone asks you, "Which of these fractions is bigger, 3/5 or 5/8?" can you answer at a glance? Most people can't. But converting to decimals makes the comparison easy:

$$3/5 = 0.6$$

and

$$5/8 = 0.625$$

Obviously, the decimal number 0.625 exceeds the decimal number 0.6, so the fraction 5/8 exceeds the fraction 3/5.

Percentages always constitute the numerators of fractions with 100 as the denominator. You write the numerator as a number followed by the word "percent" or the symbol %. You should get used to seeing either the word or the symbol. Percentages, like decimals, make it easy to compare fractions at a glance.

TIP ***When you encounter a fractional expression—something with a numerator and a denominator—you can find the percentage with a calculator. Divide the fraction out, and the calculator will "automatically" display it as a decimal expression. Then multiply that decimal number by 100 to get the figure as a percentage.***

Still Struggling

When a percentage expresses a change, it always refers to the *initial* or *starting* value. If you say that a plant grew by 1 inch, you don't know whether to call that growth large or small, unless you know how tall the thing was at the beginning. If a plant grows 1 inch taller from a starting height of 4 inches, then the percentage increase equals 1/4, or 25/100, or 25 percent. If a tree grows 1 inch taller from a starting height of 400 inches, then the percentage increase equals 1/400, or 0.25/100, or 0.25 percent.

PROBLEM 4-4

Consider the sequence of fractions where the numerator always equals 1 less than the denominator, and we keep increasing both values as follows:

1/2, 2/3, 3/4, 4/5, 5/6, ···, 99/100, ··· 999/1000, ···

What happens to the percentage as we move along in this sequence?

SOLUTION

The percentage starts out at 50% and then gradually increases toward 100%, but never gets all the way there. We can use our calculators and

work out the fraction values as decimals, rounding off to three places to get the sequence

0.500, 0.667, 0.750, 0.800, 0.833, ···, 0.990, ···, 0.999, ···

The equivalent percentages equal 100 times these values, or

50.0%, 66.7%, 75.0%, 80.0%, 83.3%, ···, 99.0%, ···, 99.9%, ···

PROBLEM 4-5

Think about the sequence of fractions where we start with 5/5, and then repeatedly increase the numerator by 1 while leaving the denominator the same, like this:

5/5, 6/5, 7/5, 8/5, 9/5, ··· 100/5, 101/5, ···, 1000/5, 1001/5, ···

What happens to the percentage as we move out in this sequence?

SOLUTION

The percentage starts out at 100% and then increases indefinitely at a steady rate. We can engage our calculators and work out the decimals to two places, getting

1.00, 1.20, 1.40, 1.60, 1.80, ···, 20.00, 20.20, ···, 200.00, 200.20, ···

To get the equivalent percentages, we multiply each of the above numbers by 100 and then rewrite the list as

100%, 120%, 140%, 160%, 180%, ···, 2000%, 2020%, ···, 20,000%, 20,020%, ···

Percentages with Money

When you deal with increasing or decreasing amounts of money, you'll hear about percentages. If railway or airline fares go up, for example, the carriers might state the change as a percentage. Dividends on stocks and other investments get paid out on a percentage basis so everything divides up fairly among the stockholders.

Imagine that a commuter railway company must raise fares because its operating costs have increased. It would not be fair to charge everyone $10.00 more. That action would raise a one-time $2.00 fare to $12.00, an increase of

$$\begin{aligned} &\$10.00/\ \$2.00 \\ &= 1000/200 \\ &= 500/100 \\ &= 500\% \end{aligned}$$

But it would raise a $200.00 full-year pass to $210.00, a much smaller percentage increase that works out to

$$\begin{aligned} &\$10.00/\$200.00 \\ &= 10/200 \\ &= 5/100 \\ &= 5\% \end{aligned}$$

Consider another example. Suppose that the profits for a company add up to $500,000 in a given year, and the company has 500 shareholders. If each shareholder gets $1000, some of them will get treated unfairly! One shareholder might have invested $100 in the company, while another one invested $100,000. If they both got $1000 of the profit for the year, the more generous shareholder would have a legitimate complaint.

If a railway's costs rise from $100,000,000 in a given year to $105,000,000 in the next year, that's a 5% increase. To get this money back in fares charged, each fare should increase by 5%. In that case, the $2.00 fare would increase to $2.00 × 1.05 = $2.10, and the $200.00 full-year pass would increase to $200.00 × 1.05 = $210.00. Similarly, if profits are $500,000 on a total investment of $10,000,000 in a given year and the dividend rate equals 5%, then the stockholder who invested $100 gets $100 × 0.05 = $5.00, and the one who invested $100,000 gets $100,000 × 0.05 = $5000.00.

TIP ***Note that in the foregoing discussion, some of the figures include a decimal point followed by two zeros (to indicate a certain number of dollars plus zero cents), while other figures don't have this extra information. You should get used to seeing money figures in both formats.***

PROBLEM 4-6

Suppose that you buy a five-year certificate of deposit (CD) at your local bank for $1000.00, and it earns interest at the annualized rate of exactly 2% per year. The bank rounds the value off to the nearest penny (cent) at the end of each year, and pays the interest then. How much will the CD be worth at the conclusion of the five-year term?

SOLUTION

The CD will have a face value of $1104.08 after five years. To calculate it, you multiply $1000.00 by 1.02, then you multiply that result by 1.02, and you repeat the process a total of five times, rounding off to the nearest penny each time.

- **After one year: $1000.00 × 1.02 = $1020.00**
- **After two years: $1020.00 × 1.02 = $1040.40**
- **After three years: $1040.40 × 1.02 = $1061.21**
- **After four years: $1061.21 × 1.02 = $1082.43**
- **After five years: $1082.43 × 1.02 = $1104.08**

Percentages Up or Down

Imagine that a woman buys a home for $400,000. After a few years its value increases, and she sells it for $500,000. She's made a 25% profit on the deal. It cost her $400,000; she recovered that amount and made $100,000 more. The gain equals 1/4 or 25/100 of the original $400,000. (For simplicity in this calculation, let's neglect fees such as the closing costs and realtor's commission.)

Now imagine a luckless man who buys the home at the $500,000 price. The value decreases, and when he loses his job, he sells the home for $400,000, the same price that the previous owner had paid for it. Having gone back to its original price, after going up 25% for the woman, does the man's misfortune represent a 25% loss? No! He paid $500,000, but he gets $400,000 back on his investment. The loss equals $100,000, the same as the profit the first owner made. But now, that $100,000 constitutes a part of $500,000, not $400,000. The loss equals 20%, which represents 1/5 or 20/100 of the original $500,000.

Does this twist seem odd to you? How can a 20% price decrease represent the same fractional change, when turned around, as a 25% price increase did for the very same property? The "discrepancy" in cases like this depends on the extent of the change. Smaller percentage changes turn out more nearly the same, either way. For larger percentages, you'll see a bigger difference.

Still Struggling

There's no limit to the percentage by which a quantity can increase. You can, at least in theory, have something increase by a million percent! But a decrease can never work out to equal more than 100%, which represents a decrease from some defined starting value all the way down to zero. You can add any imaginable multiple of something's original value to that original value, but you can only take the entire original value away once (unless you want to go into debt, pay fines, or worst of all, get the tax authorities angry at you).

PROBLEM 4-7

Compare the percentage changes of doubling a bank account (increasing it to twice the original balance) versus halving an account (shrinking it to half the original balance).

SOLUTION

Doubling the starting balance represents an increase of 100%. The whole original amount (1 = 100/100 = 100%) gets added in to our starting balance. But the reverse process, the halving, constitutes a decrease of only 50%. We take half (1/2 = 50/100 = 50%) of the original balance away.

PROBLEM 4-8

Compare the percentage change that we get when we increase the balance in a bank account by an amount equal to itself, 10 times over, to the percentage change that we get when we shrink an account down to 1/10 of the original balance.

SOLUTION

If the initial balance increases by an amount equal to 10 times itself (say from $500 to $5500), that's an increase of 1000% because we add in the original amount 10 times (10 = 1000/100 = 1000%). But if the balance goes down to only 1/10 of the original amount (say from $5500 to $550), it's a decrease of 90% because we take away 9/10 of the original (9/10 = 90/100 = 90%).

Rational or Irrational?

Mathematicians use the term *rational number* to describe a quantity that equals the ratio of an integer to a (usually different) positive integer. In this context, "rational" means "expressible as a ratio," not "sane" or "sensible."

Imagine that we assign all the rational numbers to points on a line, in such a way that the distance of any point from the origin (zero point or "center" of the line) is directly proportional to its numerical value. If a point lies on the left-hand side of the point representing 0, then that point corresponds to a negative number. If a point lies on the right-hand side of 0, it corresponds to a positive number. Figure 4-1 shows such a number line.

TIP ***All of the integers are rational numbers. You can simply divide the integer by 1 and get a fraction of the proper form for a rational number.***

TIP ***Whenever you add, subtract, or multiply one rational number by another, you get a new rational number. Whenever you divide a rational number by any other rational number except 0, you get a new rational number.***

If we take any two points on the line that correspond to rational numbers, then the point midway between them corresponds to the *average* of those two numbers. That average, technically called the *arithmetic mean*, always works out as another rational number. This rule remains true no matter how many times we repeat the operation.

We can keep cutting an interval on the *rational-number line* in half indefinitely, and if the end points both represent rational numbers, then the midpoint will always represent another rational number. Figure 4-2 shows an example of how this principle works.

- We start with the interval from 1-1/2 to 2 (or 6/4 to 8/4), cutting it in half to find a midpoint at 1-3/4 (or 7/4). Then we take the left-hand side representing the interval from 1-1/2 to 1-3/4 (or 6/4 to 7/4), and forget about the right-hand side.

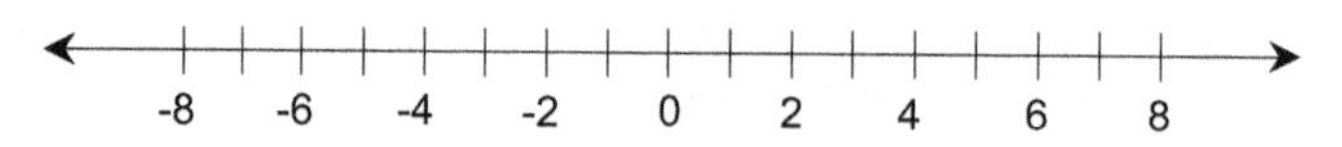

FIGURE 4-1 • We can denote the rational numbers along a straight line.

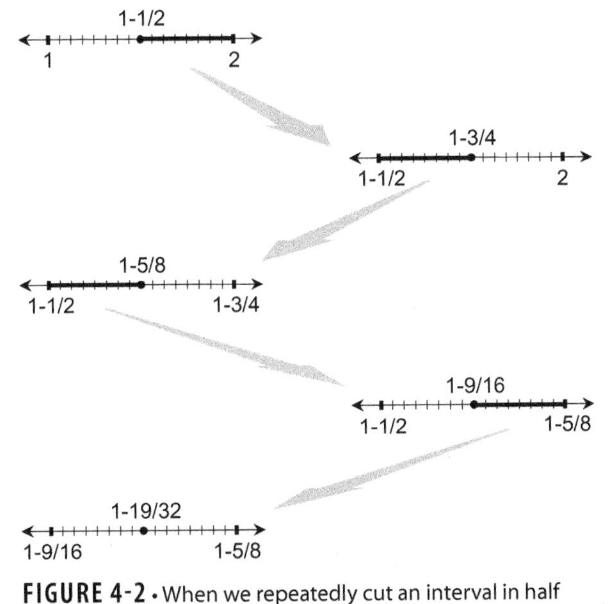

FIGURE 4-2 • When we repeatedly cut an interval in half on the rational number line, we keep coming up with new rational numbers.

- Next, we cut the interval from 1-1/2 to 1-3/4 (or 12/8 to 14/8) in half to find a midpoint of 1-5/8 (or 13/8). Then we take the left-hand side representing the interval from 1-1/2 to 1-5/8 (or 12/8 to 13/8), and forget about the right-hand side.
- Next, we cut the interval from 1-1/2 to 1-5/8 (or 24/16 to 26/16) in half to find a midpoint at 1-9/16 (or 25/16). Then we take the right-hand side representing the interval from 1-9/16 to 1-5/8 (or 25/16 to 26/16), and forget about the left-hand side.
- Finally, we cut the interval from 1-9/16 to 1-5/8 (or 50/32 to 52/32) in half, getting the midpoint that corresponds to 1-19/32 (or 51/32).
- We can keep on slicing up intervals and choosing either the left-hand half or the right-hand half, getting "messier and messier" rational numbers (but rational numbers, nevertheless) every time we get a new midpoint.

TIP ***If you've forgotten how to calculate the average of two numbers, here's the rule: Add the numbers and then divide the result by 2. Take signs into account when you work with negative numbers. For example, the average of 2 and 8 is***

$$(2 + 8)/2 = 10/2$$
$$= 5$$

but the average of 2 and −8 is

$$[2 + (-8)]/2 = (2 - 8)/2$$
$$= -6/2$$
$$= -3$$

Still Struggling

Do you imagine that points on a line, defined as corresponding to rational numbers, are "infinitely dense"? They are clustered so closely together that every point in between any two arbitrary rational-number points corresponds to another rational number. But do the rational numbers account for *all* of the points along a true geometric line? The answer, which surprises many people the first time they hear it, is "No."

PROBLEM 4-9

Suppose that we come across a decimal expression in which a certain pattern repeats endlessly to the right of the decimal point. Does this expression represent a rational number? If so, why? If not, why not?

SOLUTION

Yes, it does. As we learned in the last chapter, whenever we see a repeating pattern of digits to the right of a decimal point, and if that pattern repeats "forever" (no matter how far out to the right we keep writing the digits), then we can represent the quantity to the right of the decimal point as the sequence of digits divided by an equal number of 9s. Here are some examples:

$$0.444444444444 \cdots = 4/9$$

$$0.232323232323 \cdots = 23/99$$

$$-0.861861861861 \cdots = -861/999$$

$$0.238723872387 \cdots = 2387/9999$$

If we find anything other than 0 to the left of the decimal point when we see a repeating pattern to the right of the point, we still have a rational

number. So, for example, the following expressions all represent rational numbers:

$$1.444444444444\cdots$$

$$4.232323232323\cdots$$

$$-8.861861861861\cdots$$

$$789.238723872387\cdots$$

An *irrational number* can't be expressed as the ratio of two integers. Examples of irrational numbers include:

- The length of the diagonal of a square that measures exactly 1 unit along each edge
- The length of the diagonal of a cube that measures exactly 1 unit along each edge
- The circumference-to-diameter ratio of any circle
- The surface-area-to-radius ratio of any sphere

All irrational numbers share the property of being inexpressible in decimal form. When we try to do it, we get a *nonterminating, nonrepeating* decimal expression, as we saw in the last chapter. No matter how many digits we write down to the right of the decimal point, we can only get an approximation of the actual value of the number. We can't find any pattern, so we can't even denote the number as an "implied endless list of digits," such as the ones shown above for rational numbers.

TIP *No irrational number is rational, and no rational number is irrational. The two collections compose entirely separate things. A pure mathematician would call them* disjoint sets.

TIP *In the context of numbers, the term "irrational" simply tells you that it's impossible to express a quantity as a ratio of an integer to a positive integer. It has nothing to do with "sensibility" or "sanity."*

Real Numbers

When we combine all the rational numbers and all the irrational numbers into a single massive collection, we get the *real numbers*. We can denote the real numbers on a straight line that runs off endlessly in both directions, just as we did with the rational numbers in Fig. 4-1. When we glance at it casually, the real number line looks like the rational number line. But when we scrutinize the real number line "under a mathematical microscope," we find that it *drastically differs* from the rational number line!

We've seen that the rational numbers, when depicted as points along a line, are "dense." No matter how close together two rational-number points might happen to lie, we can always find another rational-number point between them. But this fact doesn't give us permission to conclude that the rational numbers account for *all* of the points that can exist on a number line. We can assign points to irrational numbers along the same line as we assign points to rational numbers. When we do that, we get a more "dense" line. It's as if we took a thread made of spun cotton dyed a coppery color and turned it into a wire made of real copper. The thread and the wire might look the same to a casual observer, but when we compare their densities, we discover the difference.

Still Struggling

You might ask, "How many times more dense is the real-number line than the rational-number line? Twice? A dozen times? A hundred times?" As things work out, the collection of real numbers, when depicted as points on a line, is *infinitely* more dense than the collection of rational numbers. If you have trouble envisioning such a thing, you're not alone. A proof of this fact would take us beyond the scope of this book, but mathematicians have done it. You might imagine the situation as follows: Even if you could live to be "infinity years old," and even if you spent all your time naming real numbers one by one, you'd never live long enough to see the year in which you could claim to have named them all.

PROBLEM 4-10

Think about two integers so large that it would take us millions of years to write either of them out by hand. We are told that both of these integers are prime numbers, and that the decimal expansion of their ratio produces an endless sequence of digits. Is there a repeating pattern to the digits in the decimal expansion?

SOLUTION

This decimal expansion—an endless string of digits—has a repeating pattern. The original quotient constitutes a rational number by definition. Remember, *any* rational number can be expressed as either a terminating decimal or an endless repeating decimal. The repeating pattern of digits in the decimal expansion might turn out incredibly long, but it's finite. If we have unlimited time and if we have endless lives, we'll find the pattern sooner or later!

PROBLEM 4-11

Imagine that we come across a gigantic string of digits, hundreds of kilometers long if we try to write it out, and we can't identify any pattern. We use a computer to examine this number to a thousand decimal places, then a million, then a billion, and we still can't find a pattern. Can we ever know for sure whether or not a pattern exists, so we can decide whether or not the number is rational?

SOLUTION

Trouble often comes up when we try to force hard reality into the mold of pure theory. Even the most powerful supercomputer can get confused by a string-of-digits problem, if the repeating pattern is complicated enough. But the fact that a pattern can't be discovered in a human lifetime does not prove conclusively that no pattern exists! It works the other way, too. If we see a long string of digits repeating many times, we can't be sure that it will repeat endlessly, unless we know that there's a ratio of integers with the same value.

QUIZ

Refer to the text in this chapter if necessary. A good score is eight correct. Answers are in the back of the book.

1. The fraction 15/10 is larger than the counting number 1 by an amount equal to one of the following percentages (rounded off to the nearest whole-number value). Which percentage?

A. 100%
B. 67%
C. 50%
D. 33%

2. The counting number 1 is a certain percentage of 15/10 (rounded off to the nearest whole-number value). Which percentage?

A. 100%
B. 67%
C. 50%
D. 33%

3. Consider the following sequences of ratios. In one, but only one, of these sequences, the right-hand quantity varies in inverse proportion to the left-hand quantity. Which sequence is it?

A. 12:6, 13:7, 14:8, 15:9, 16:10, 17:11
B. 243:1, 81:3, 27:9, 9:27, 3:81, 1:243
C. 18:6, 15:5, 12:4, 9:3, 6:2, 3:1
D. 1:1, 2:2, 3:3, 4:4, 5:5, 6:6

4. Suppose that you buy a 10-year certificate of deposit (CD) at your local bank for $5000.00, and it earns interest at the annualized rate of exactly 3% per year. The bank rounds the value off to the nearest penny at the end of each year, and pays interest then and only then. How much will the CD be worth at the conclusion of the 10-year term?

A. $6719.59
B. $6105.55
C. $5876.04
D. $5543.43

5. A home-heating-system manufacturer tells you that a Model X furnace can provide 80,000 British thermal units per hour (Btu/h) of heating power for your home, while a Model Y furnace can provide 100,000 Btu/h. What's the ratio of the Model Y furnace's output to the Model X furnace's output?

A. 100:80
B. 10:8
C. 5:4
D. All of the above

6. In the scenario of Question 5, what's the ratio of the Model X furnace's output to the Model Y furnace's output?

A. 1:1.2
B. 1:1.25
C. 1:1.33
D. 1:1.4

7. Someone tells you that she has two bank accounts. Her Happyville Bank account has $344.66 in it, and her Bluesdale Bank account contains $2757.28. You can correctly tell her that the ratio of her Bluesdale balance to her Happyville balance equals

A. 1:8.
B. 8:1.
C. 34,466 / 275,728.
D. Any of the above

8. When we want to multiply one fraction by another fraction, we multiply their numerators to get the new numerator, and we multiply their denominators to get the new denominator. Knowing this rule, we can show that the product of the rational numbers 0.555555··· and 0.777777··· is another rational number because it equals

A. $(5/9) + (7/9)$, which equals $12/9$.
B. $(5/9) \times (7/9)$, which equals $35/81$.
C. $(5/9) \times (9/7)$, which equals $45/63$.
D. Any of the above

9. When we want to divide one fraction by another fraction, we "invert" the second fraction (transpose the numerator and denominator), and then we multiply the resulting two fractions by each other. Knowing this rule, we can show that the ratio of the rational number 0.567567567··· to the rational number 0.765765765··· is another rational number because it equals

A. $(567/999) + (765/999)$, which equals $1332/999$.
B. $(567/999) \times (765/999)$, which equals $433{,}755/998{,}001$.
C. $(567/999) \times (999/765)$, which equals $566{,}433/764{,}235$.
D. None of the above

10. Consider the following sequences of ratios. In one, but only one, of these sequences, the right-hand quantity varies in direct proportion to the left-hand quantity. Which sequence is it?

A. 12:6, 13:7, 14:8, 15:9, 16:10, 17:11
B. 243:1, 81:3, 27:9, 9:27, 3:81, 1:243
C. 18:6, 15:5, 12:4, 9:3, 6:2, 3:1
D. 2:3, 3:4, 4:5, 5:6, 6:7, 7:8

chapter 5

Operations

Have you ever visited your community bank and needed your calculator, but forgot to bring it? In situations like that, you can do basic arithmetic "by hand." All you need is a pencil or pen, and a little bit of scratch paper. You learned how to do "longhand arithmetic" in grammar school. If it has faded in your memory, or even if you've forgotten it altogether, let's bring it back!

CHAPTER OBJECTIVES

In this chapter, you will

- Analyze the mechanics of longhand addition.
- Learn how to carry digits.
- Analyze the mechanics of longhand subtraction.
- Learn how to borrow digits.
- Analyze the mechanics of longhand multiplication.
- Analyze the mechanics of longhand division.
- Convert longhand division remainders into fractions.

Longhand Addition

You probably memorized all the single-digit addition facts in grammar school to the point where you could "do them like a computer." Even today, you might know instantly, without having to think, that $5 + 8 = 13$ or that $9 + 8 = 17$.

Table 5-1 shows the addition facts for single-digit numbers. You can add multiple-digit numbers by combining ones with ones, tens with tens, hundreds with hundreds, and so on. Just as $2 + 3 = 5$, for example, you can add larger numbers, such as the following:

$$30 + 20 = 50$$

$$300 + 200 = 500$$

$$33 + 22 = 55$$

$$333 + 222 = 555$$

PROBLEM 5-1

Add 312 and 406, starting with the ones place, then going to the tens place, then finishing up with the hundreds place. Arrange all the individual digits in an

TABLE 5-1 Addition facts for single digits. Pick a number in the leftmost column, add it to the number in the topmost row, and read the sum in the box where the column and row intersect.

	0	**1**	**2**	**3**	**4**	**5**	**6**	**7**	**8**	**9**
0	0	1	2	3	4	5	6	7	8	9
1	1	2	3	4	5	6	7	8	9	10
2	2	3	4	5	6	7	8	9	10	11
3	3	4	5	6	7	8	9	10	11	12
4	4	5	6	7	8	9	10	11	12	13
5	5	6	7	8	9	10	11	12	13	14
6	6	7	8	9	10	11	12	13	14	15
7	7	8	9	10	11	12	13	14	15	16
8	8	9	10	11	12	13	14	15	16	17
9	9	10	11	12	13	14	15	16	17	18

array with rows and columns (also called a *matrix*) with the ones in the right-hand column, the tens to the left of that, and the hundreds to the left of that.

SOLUTION

Take the ones first, adding 2 + 6 to get 8. Next take the tens, adding 1 + 0 to get 1. Last take the hundreds, adding 3 + 4 to get 7. You end up with seven hundreds, one ten, and eight ones, which total 718. Arrange the digits in a matrix with the ones in the far-right column, the tens in the column to the left of that, and the hundreds in the column to the left of that. You can place a plus sign (+) in front of the second quantity to indicate that you should add. The third row, with the long horizontal line, separates the top two figures (the *addends*) from the result (the *sum*). Here's what you get:

$$\begin{array}{rrr} 3 & 1 & 2 \\ +4 & 0 & 6 \\ \hline 7 & 1 & 8 \end{array}$$

TIP *Problem 5-1 has numbers in each place that don't add up to over 9. If any number group or place adds up to 10 or more, you must move the extra digit 1 (which actually represents 10) to the next higher group or place. The next three problem/solution pairs illustrate how this process works.*

PROBLEM 5-2

Add 746 and 158, starting with the ones place, then going to the tens place, then finishing up with the hundreds place.

SOLUTION

First, add the quantities in the ones place: 6 + 8 gives you 14. In this sum, the numeral 1 belongs in the tens place. Instead of only 4 and 5 to add in the tens place, you have an extra 1 from adding 6 + 8. You move, or *carry*, that extra digit to the top of the tens (middle) column. The same thing will happen whenever the sum at a certain place exceeds 9. In the tens place, add 1 + 4 + 5 to get 10. Now you have 0 in the sum (bottom) row for the tens (middle) column, so you must carry the extra 1 to the hundreds place (top of the leftmost column). You add 1 + 7 + 1 to get 9 for the hundreds place, finishing up with nine hundreds, zero tens, and four ones, which equals 904. You can illustrate this process with a matrix similar to the one in the solution to Problem 5-1, adding an extra row at the top to show the carried digits (numerals 1 in small type), as follows:

1	1	
7	4	6
+1	5	8
9	0	4

PROBLEM 5-3

Add 6749 and 7585, starting with the ones place, then going to the tens place, then going on to the hundreds and thousands places, and finally reaching the ten-thousands place.

SOLUTION

Begin by adding 9 + 5 in the ones column, getting 14. Then carry the 1 to the top of the tens column. After that, proceed as follows:

- Add 1 + 4 + 8 to get 13 in the tens column
- Leave the 3 in the tens column
- Carry the 1 to the top of the hundreds column
- Add 1 + 7 + 5 to get 13 in the hundreds column
- Leave the 3 in the hundreds column
- Carry the 1 to the top of the thousands column
- Add 1 + 6 + 7 to get 14 in the thousands column
- Carry the 1 to the top of the ten-thousands column

Because no other digits exist in the ten-thousands column, you bring the carried digit 1 down to the bottom of that column. Here's how the worked-out matrix looks:

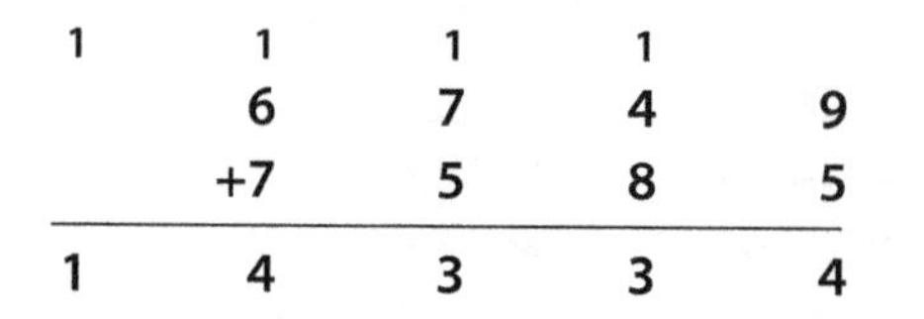

1	1	1	1	
	6	7	4	9
	+7	5	8	5
1	4	3	3	4

If the blank spots in the leftmost column cause confusion, you can place ciphers (digits 0) in those spots and then add 1 + 0 + 0 in that column to get 1, as follows:

1	1	1	1	
0	6	7	4	9
+0	7	5	8	5
1	4	3	3	4

TIP *Note that the bottom rows in the preceding two matrices say 14334, not 14,334. When you work with addition matrices, you don't have to put commas in rows that contain numerals larger than 9999, as you'd normally do when writing such numerals out in plain text.*

TIP *When you want to add decimal quantities using arrays such as the ones above, arrange the quantities so that all the decimal points lie in a single vertical column. Let all the decimal points "add up" to another decimal point in the bottom row. Then add the digits just as you would do with whole numbers. Unlike the situation with commas, you absolutely must include decimal points in your addition arrays!*

PROBLEM 5-4

Work out the dollar-and-cent sum \$59.99 + \$287.87 + \$1078.78, taking advantage of the matrix method and carrying techniques used in the previous three examples. Work from right to left, starting with pennies (cents), then going to dimes (units of 10 cents), then to dollars, tens of dollars, hundreds of dollars, and thousands of dollars in that order.

SOLUTION

You must arrange the numbers so that all the decimal points line up vertically in their own column, and they all "add up" to another decimal point at the bottom. When you do that, you get the following partially filled-in array:

		5	9	.	9	9
	+2	8	7	.	8	7
+1	0	7	8	.	7	8
		(Answer will go here)				

First, add 9 + 7 + 8 in the pennies column to get 24. Carry the 2 to the top of the dimes column. After that, continue from right to left, column by column, as follows:

- Add 2 + 9 + 8 + 7 to get 26 in the dimes column
- Leave the 6 in the dimes column
- Leave the decimal-point column alone

- **Carry the 2 to the top of the dollars column**
- **Add 2 + 9 + 7 + 8 to get 26 in the dollars column**
- **Leave the 6 in the dollars column**
- **Carry the 2 to the top of the tens-of-dollars column**
- **Add 2 + 5 + 8 + 7 to get 22 in the tens-of-dollars column**
- **Leave the ones digit 2 in the quantity 22 in the tens-of-dollars column**
- **Carry the tens digit 2 in the quantity 22 to the top of the hundreds-of-dollars column**
- **Add 2 + 2 + 0 to get 4 in the hundreds-of-dollars column**
- **Bring down the 1 to the bottom of the thousands-of-dollars column**

You end up with one thousand, four hundred twenty-six dollars and sixty-four cents. The process produces the following matrix:

	2	2	2		2	
		5	9	.	9	9
	+2	8	7	.	8	7
+1	0	7	8	.	7	8
1	4	2	6	.	6	4

TIP ***Bookkeepers often use two methods to add long columns of numbers. First they add from the top down, and then they add from the bottom up. By following that redundant scheme, they can check their work. If you've done much accounting, then you know how easily mistakes can creep into long sums!***

Still Struggling

You can add two or more quantities in any order you want, and you'll always get the same result (as long as you don't make any mistakes). Mathematicians call this rule the *commutative law for addition*.

Longhand Subtraction

Many people have trouble with subtraction. You can memorize the subtraction facts for all the single-digit numbers. Then you'll know instantly, for example, that $8 - 2 = 6$ or that $7 - 5 = 2$ without having to "count away."

You can subtract multiple-digit numbers by combining ones with ones, tens with tens, hundreds with hundreds, and so on, in the same way as you do with addition. Then you work on each column as its own "subtraction subproblem." Here are some examples:

$$30 - 20 = 10$$

$$300 - 200 = 100$$

$$33 - 22 = 11$$

$$333 - 222 = 111$$

PROBLEM 5-5

Subtract 153 from 876, starting with the ones place, then going to the tens place, then finishing up with the hundreds place. In other words, find the *difference* 876 − 153.

SOLUTION

Take the ones first, subtracting 6 − 3 to get 3. Next take the tens, subtracting 7 − 5 to get 2. Last take the hundreds, subtracting 8 − 1 to get 7. You end up with seven hundreds, two tens, and three ones, which total 723. You can arrange the single digits in a matrix, just as you did with addition, but with a minus sign in front of the second numeral (telling you to subtract), as follows:

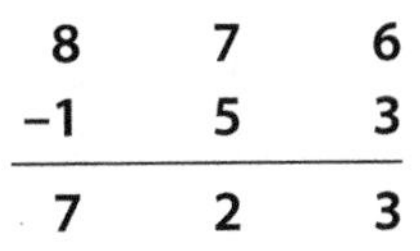

8	7	6
−1	5	3
7	2	3

TIP ***When you work out subtraction problems, instead of carrying digits from one column to another, you can "steal" values from one column and give them to another if you get a negative number in any particular column.***

PROBLEM 5-6

Suppose that you want to subtract 129 from 256. In other words, you want to find the difference between the two numbers, written as 256 − 129. How can you accomplish this task without getting negative numbers in any of the matrix "cells"?

SOLUTION

First, subtract the numbers in the ones column. If you do that directly, you get 6 − 9, which equals −3. That's a perfectly good numerical value, but it

doesn't make sense in the ones place of a multiple-digit numeral! You can deal with this problem by "raiding" a one digit value from the tens column (the equivalent of 10) and adding that 10 to the ones column. Then you get 6 + 10, or 16, at the top of the ones column, and you reduce the value at the top of the tens column from 5 to 4. The following matrix shows how the process goes:

	4	16
2	~~5~~	~~6~~
−1	2	9
1	2	7

You cross out the "old" values and write the new ones above them. Now you have 16 − 9 in the ones column, which leaves 7 there, and 4 − 2 in the tens column, which leaves 2 there. You don't have to do anything special to the hundreds column; you can simply subtract 2 − 1 to get 1 in the bottom row there. Now you have the answer: one hundred, two tens, and seven ones, which comes out to 127.

You can check the foregoing result by *adding back* (from the bottom up in the subtraction matrix), summing 127 + 129. If you did all your work correctly, you should get 256. Here's the addition matrix:

	1	
1	2	7
+1	2	9
2	5	6

TIP *When you check subtraction by adding back, you get extra assurance that you've subtracted correctly. If you don't get the original number at the end, you'd better work the problem out again!*

PROBLEM 5-7

Use longhand subtraction to calculate the difference 27,214 − 18,538. You'll have to borrow several times.

SOLUTION

The matrix below displays the process. You can break the process down into neat steps. Follow along in the array as you read through the process, one step at a time:

- Start with the ones place, where you must take 8 from 4. You can't "legally" do that, so you have to borrow a 1 from the tens place to get 14 – 8, which works out to 6. That's the ones digit in your answer.
- Now move to the tens place, where you have 0 – 3. Again, you get a negative number if you take that difference directly, so you have to borrow a 1 from the hundreds place to get 10 – 3, which leaves 7 in the tens place for the answer.
- Moving on to the hundreds place, you have to find 1 – 5, which again works out negative, so you borrow a 1 from the thousands place. Subtracting 11 – 5 in the hundreds place leaves 6 as your answer there.
- Going to the thousands place, you encounter the difference 6 – 8, so you have to borrow yet again, this time from the ten-thousands place. When you do that, you have 16 – 8 in the thousands place, leaving 8 as your answer there.
- Finally you get to the ten-thousands place, where 1 – 1 = 0. You don't have to write the 0 in the bottom row because it's the leftmost digit in a multiple-digit numeral.
- You've worked out the fact that 27,214 – 18,538 = 8676.

	16	11	10	
1	~~6~~	~~1~~	~~0~~	14
~~2~~	~~7~~	~~2~~	~~1~~	4
–1	8	5	3	8
	8	6	7	6

Now you should check your work by adding back! You must calculate the sum 8676 + 18,538, and if you did your subtraction properly, you'll end up with 27,214. Here's the matrix for that process, which you should be able to follow without a detailed description:

1	1	1	1	
	8	6	7	6
+1	8	5	3	8
2	7	2	1	4

PROBLEM 5-8

Imagine that you cash a check at your bank in the amount of \$388.89. Your ledger shows a balance of \$1003.66 to start out with. You forgot to bring your calculator, so you do the subtraction manually to keep your account current. What's the amount remaining in the ledger after you've cashed the check?

SOLUTION

As you did in the previous problem, you can arrange the values in a matrix, this time adding a special column for the decimal point that separates dollars from cents. Then you work the problem out column by column starting with the pennies and going to the left. Refer to the matrix below as you follow along:

- You start with the pennies place, where you must take 9 from 6. You borrow a 1 from the dimes place to get 16 – 9, which leaves you with 7. That's the pennies digit in your answer.
- You move to the dimes place, where you have 5 – 8. You borrow a 1 from the dollars place to get 15 – 8, which leaves 7 in the dimes place for your answer.
- Moving on to the dollars place, you have to find 2 – 8, so you decide that you need to borrow a 1 from the tens-of-dollars place to get 12 in the dollars place.
- Now you're stuck! How can you borrow when you have 0 in a column to start out with? If you look at the three digits in the thousands-of-dollars, hundreds-of-dollars, and tens-of-dollars columns combined, you see 100. You can borrow one from that entire quantity to get 99, with one 9 in the hundreds-of-dollars place and one 9 in the tens-of-dollars place. The digit 1 in the thousands-of-dollars place turns into 0. (You don't have to put down that 0 in the top row because it's in the leftmost column, so in effect, it disappears.)
- Now you've got 12 at the top of the dollars column, having borrowed 1 from the 100 figure in the three columns to its left.
- Subtracting 12 – 8 leaves you with 4 as your result in the dollars column.
- Now going to the tens-of-dollars column, you have 9 – 8, which leaves 1 as your answer there.
- In the hundreds-of-dollars column, you have 9 – 3, leaving 6 as your answer there.
- You've just updated your ledger by calculating \$1003.66 – \$388.89 = \$614.77.

			12		15	
	9	9	~~2~~		~~5~~	16
~~1~~	~~0~~	~~0~~	~~3~~	.	~~6~~	~~6~~
	–3	8	8	.	8	9
	6	1	4	.	7	7

You don't want to have an inaccurate account balance in your records, so you'd better check your work by adding back. When you do that, you write down the following array of digits:

1	1	1	1		1	
	6	1	4	.	7	7
	+3	8	8	.	8	9
1	0	0	3	.	6	6

Still Struggling

You can't subtract numbers in reverse order and get the same result either way. When you reverse the order of a difference, you get the *negative* (also called the *additive inverse*) of the answer you got before. The negative of a number equals −1 times that number. A quantity and its negative always add up to 0. You'll get the same result either way only when both starting values are equal. A mathematician would say that subtraction is *anticommutative*.

Longhand Multiplication

Imagine that you stroll into your local "dollar store" and buy six articles for \$2.00 each. The total cost (before tax) comes to \$12.00; you count \$2.00 six times. What if the articles cost \$5.00 apiece? In that case, to find the total, you count \$5.00 six times to get \$30.00. This sort of repeated addition constitutes *multiplication*.

When you attended grammar school, did your teachers make you memorize multiplication tables? If you memorized, say, 6 times 2 equal 12 (written $6 \times 2 = 12$) or six times 5 equal 30 (written $6 \times 5 = 30$), then you could do simple calculations without having to use a table or calculator. Table 5-2 portrays the single-digit multiplication facts up through $9 \times 9 = 81$.

As a child, you learned the matrix method for multiplying multiple-digit numbers by single-digit numbers. For example, take 123×2. Let's review that

TABLE 5-2 Multiplication facts for single digits. Pick a number in the leftmost column, multiply it by the number in the topmost row, and read the product in the box where the column and row intersect.

	0	**1**	**2**	**3**	**4**	**5**	**6**	**7**	**8**	**9**
0	0	0	0	0	0	0	0	0	0	0
1	0	1	2	3	4	5	6	7	8	9
2	0	2	4	6	8	10	12	14	16	18
3	0	3	6	9	12	15	18	21	24	27
4	0	4	8	12	16	20	24	28	32	36
5	0	5	10	15	20	25	30	35	40	45
6	0	6	12	18	24	30	36	42	48	54
7	0	7	14	21	28	35	42	49	56	63
8	0	8	16	24	32	40	48	56	64	72
9	0	9	18	27	36	45	54	63	72	81

system. You place the digits in an array with the larger number on top and the single-digit number underneath, like this:

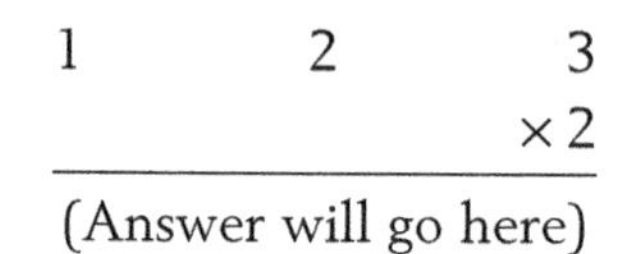

First, you multiply 2 × 3, getting 6, which you put down in the ones place of the answer row (the bottom row, under the horizontal line). Then you multiply 2 × 2 to get 4, which you write in the tens place of the answer row. You finish up by taking 2 × 1 to get 2, which you write in the hundreds place of the answer row. You end up with the following matrix:

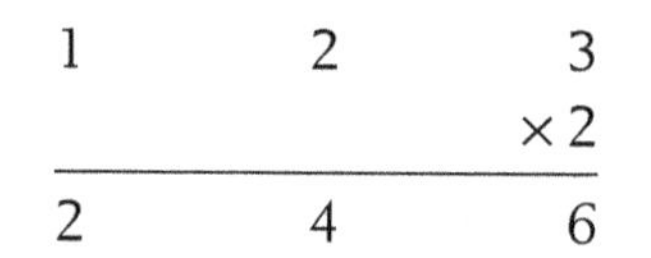

You can carry digits in multiplication when you get double-digit quantities at any stage. Consider 147 × 5. Set up the problem by arranging the digits like this:

$$\begin{array}{ccr} 1 & 4 & 7 \\ & & \times 5 \\ \hline \end{array}$$

(Answer will go here)

First, multiply 5 × 7 to get 35. Leave the 5 in the ones place for your answer, and carry the 3 to the tens place. Then multiply 5 × 4 to get 20, adding the carried 3 to get 23. Leave the 3 in the tens place for the answer, and carry the 2 to the hundreds place. Then multiply 5 × 1 to get 5, adding the carried 2 to get 7, which you bring down to the hundreds place for the answer. Now read through the foregoing steps again and follow the digits below:

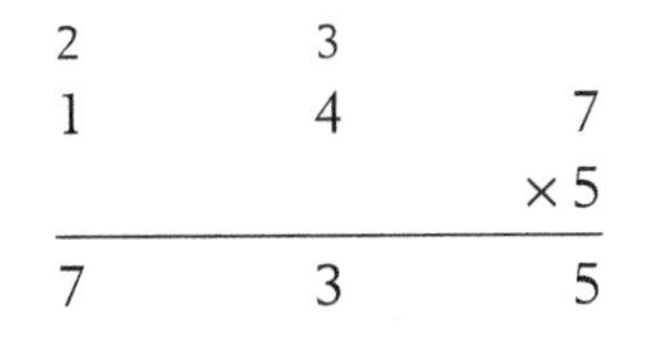

What can you do if the second number in a product has two digits? Consider, for example, 147 × 65. In this case, you multiply the top quantity by the ones digit in the second number, and then you multiply the top quantity by the tens digit in the second number. But then you have to move the second product over to the left by one place, and add the two "intermediate answers" to get your final answer. The next problem illustrates how things go.

PROBLEM 5-9

Work out 147 × 65 using the matrix method. Describe the process step-by-step and show the matrix.

SOLUTION

The process breaks down as follows. As you read through these steps, "chase the digits" in the matrix below.

- **Multiply 147 × 5 as you did in the previous example, getting 735. The carried digits 2 and 3 show up above the 1 and 4, in the second small row from the top.**
- **Multiply 147 × 6 to get 882, moving all the digits over one place to the left, so the ones digit in the "intermediate answer" lines up with the 6 in 65. The carried digits 2 and 4 show up above the 1 and the 4 in the first quantity, in the topmost small row.**
- **Insert a cipher (digit 0) after 882 to show that those digits actually stand for 8820. (When you worked out the product 147 × 6 in the previous step, you really multiplied by 60 because the digit 6 occupies a spot in the tens place. So your final product works out as 10 × 882 = 8820.)**

- Add 735 + 8820 to get 9555, the final answer. The carried digit 1 from this addition step goes in the leftmost spot in the small row immediately below the upper horizontal line.
- Read through all these steps again so that you "get the feel of the process," and also to double-check your work.
- If you want reassurance, multiply the problem out on a calculator to triple-check.

	2	4	
	2	3	
	1	4	7
		×6	5
1			
	7	3	5
+8	8	2	0
9	5	5	5

Still Struggling

You can multiply two numbers in either order and get the same result (as long as you don't make any mistakes). Mathematicians call this rule the *commutative law for multiplication*. You can take advantage of it when you set up multiplication arrays. If one of the numerals has more digits than the other, place the longer numeral (the one with more digits) in top row of the matrix.

Now suppose that the second number in a product has three digits, such as when you want to calculate 147 × 265. In this case you'll multiply the top quantity by the ones digit in the second number, then multiply the top quantity by the tens digit in the second number, and finally multiply the top quantity by the hundreds digit in the second number. You'll have to move the second product over to the left by one place, and move the third product over to the left by two places before adding up all three "intermediate answers" to obtain the final answer.

PROBLEM 5-10

Work out the product 147 × 265 using the matrix method. Describe the process and show the matrix.

SOLUTION

The following steps break down the work. As you read through it all, keep track of events in the matrix below.

- Multiply 147 × 5, just as you did in the previous two examples, getting 735. The carried digits 2 and 3 appear in the third small row from the top.
- Multiply 147 × 6 to get 882, moving all the digits over one place to the left, so the ones digit in the "intermediate answer" lines up with the 6 in 265. The carried digits 2 and 4 show up in the second small row from the top.
- Multiply 147 × 2 to obtain 294, moving all the digits two places over to the left, so the ones digit in this "intermediate answer" lines up with the 2 in 265. The carried digit 1 appears in the topmost small row.
- Insert a cipher after 882 to show that those digits actually stand for 8820, for the same reason you did it in the solution to Problem 5-9.
- Insert two ciphers after 294 to show that those digits actually stand for 29,400. (When you worked out 147 × 2, you really multiplied by 200 because the digit 2 occupies a spot in the hundreds place, so your final product is 100 × 294 = 29,400.)
- Add 735 + 8820 + 29,400 to get 38,955, the final answer. The two carried digit 1s from this addition step go in the two leftmost spots in the row just below the upper horizontal line.
- Read through all these steps again so that you "get the feel of the process," and also to double-check your work.
- If you want reassurance, multiply the problem out on a calculator.

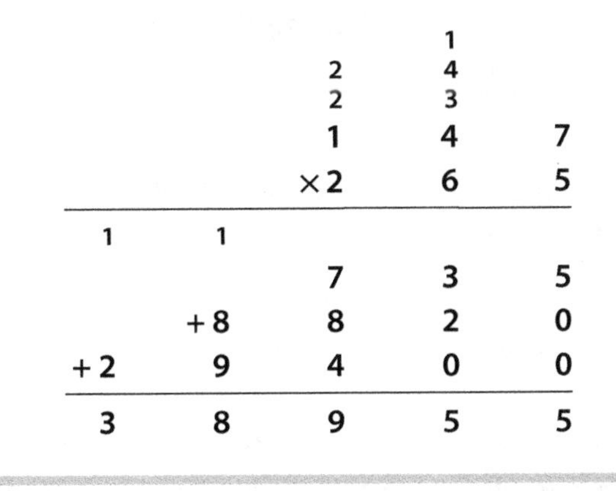

TIP *At this point, you might ask, "How does longhand multiplication work when one or both of the quantities contain decimal points?" It's easy. First, total up the number of digits to the right of the decimal point in both numerals. (For example,*

if the first quantity has one digit to the right of the decimal point and the second quantity has two digits to the right of the decimal point, that's three digits total.) Then remove the decimal points from the expressions and multiply them as if they were whole numbers. When you get your answer, count back from the right-hand end of the numeral by the total number of digits to the right of the decimal you found earlier, and put the decimal point there.

PROBLEM 5-11

Suppose that you earn $14.70 per hour at your job. Someone tells you that the president of the company earns 26.5 times as much as you do, per hour of work that she does. How much, per hour, does she earn?

SOLUTION

Note that $14.70 equals 14.7 dollars. That quantity has one digit to the right of the decimal point. The "wage factor," 26.5, also has one digit to the right of the decimal point. That's two digits, in total, that lie to the right of decimal points. Multiply 147×265 to get 38,955 (just as you did in the solution to Problem 5-10). Then count back from the right by two digits to get 389.55. The company president makes $389.55 per hour. Do you think she's worth that much?

PROBLEM 5-12

Work out the product 0.361×1.72 using the matrix method. Describe the process and show the matrix.

SOLUTION

The following steps describe the work. As you read through it, keep track of events in the matrix below.

- Count up all the digits that lie to the right of decimal points. The first quantity has three, and the second quantity has two, so that's a total of five digits.
- You can remove the 0 from the left of the decimal point in the first expression before dropping the decimal points in both expressions. Now you want to find the product 361×172.
- Multiply 361×2 to obtain 722. The carried digit 1 appears above the 3 in the quantity 361, in the second small row from the top.

- **Multiply 361 × 7 to get 2527, moving all the digits over one place to the left, so the ones digit in the "intermediate answer" lines up with the 7 in 172. The carried digit 4 shows up above the 3 in the number 361, in the small row at the very top.**
- **Multiply 361 × 1 to obtain 361, moving all the digits two places over to the left, so the ones digit in the "intermediate answer" lines up with the 1 in 172.**
- **Insert a cipher after 2527 to show that those digits actually stand for 25,270.**
- **Insert two ciphers after 361 (representing 361 × 1), to show that those digits actually stand for 36,100.**
- **Add 722 + 25,270 + 36,100 to get 62,092. The two carried digit 1s from this addition step go in the two leftmost spots in the row just below the upper horizontal line.**
- **Count back five digits from the extreme right and insert a decimal point in the preceding result, getting .62092. If you like, you can write down a cipher to the left of the decimal point to get your final answer in the format that most people prefer, 0.62092.**
- **Read through all these steps again so that you "get the feel of the process," and also to double-check your work.**
- **If you want reassurance, multiply the problem out on a calculator.**

		4		
		1		
		3	6	1
		×1	7	2
1	1			
		7	2	2
2	5	2	7	0
3	6	1	0	0
6	2	0	9	2

Longhand Division

Suppose that you visit your local grocery store and buy six identical articles for a total bill of $42.00 (before tax). You can figure out that each item cost $7.00

because $7.00 × 6 = $42.00. Because you have whole dollars, you can divide 42 by 6 to get 7, a fact that you denote as

$$42/6 = 7$$

What if the articles total up to $72.00? In that case, to find the total, you'll count $12.00 six times to get $72.00. You can divide 72 by 6 to get 12, a fact that you write down as

$$72/6 = 12$$

You can perform more complicated division problems "by hand" using a matrix system something like the ones that you've seen so far in this chapter. In a division problem, mathematicians call the answer the *quotient*.

PROBLEM 5-13

Use the matrix method for longhand division to find the quotient 3647 divided by 7.

SOLUTION

Set up the problem as shown below. The number that you want to divide up (called the *dividend*, which you can also portray as the numerator in a ratio) goes under a double horizontal line. In this case, it's 3647. The number that you want to divide by (called the *divisor*, which you can also portray as the denominator in a ratio) goes to the left of the dividend. You separate the divisor from the dividend with a closing parenthesis. In this situation, the divisor is 7. You can read the expression "7) 3647" out loud as "7 goes into 3647." It represents the ratio or fraction 3747/7.

	(Answer will go here)			
7)	3	6	4	7

The following steps describe the work. As you read through all this, keep track of events in the matrix below.

- **Figure out the largest whole-number multiple of the divisor (7) that goes into the first digit of the dividend (3). The answer is none. You can't do anything yet.**
- **Figure out the largest whole-number multiple of the divisor (7) that you can get into the quantity represented by the first two digits of the dividend (36). The answer is 5.**

- **Write down the digit 5 in the answer row above the 6 in the dividend numeral 3647. Then note that $5 \times 7 = 35$. Write down 35 below the digits 36 in 3647.**
- **Subtract 36 – 35 to get 1, and bring down the 4 from the dividend to make 14.**
- **Determine the largest whole-number multiple of 7 that you can fit into 14. The answer is 2.**
- **Write down the digit 2 after the digit 5 in the answer row. Then note that $2 \times 7 = 14$. Write down these digits 14 below the 14 that you got when you subtracted 36 – 35 and brought down the 4.**
- **Subtract 14 – 14 to get 0, and bring down the 7 from the dividend to make 07.**
- **Determine the largest whole-number multiple of 7 that you can fit into 7. The answer is 1.**
- **Write down the digit 1 after the digit 2 in the answer row. Then note that $1 \times 7 = 7$. Write down this digit 7 below the 07 that you got when you subtracted 14 – 14 and brought down the 7.**
- **Subtract 07 – 7 to get 0. This quantity 0 indicates that 521 is the exact answer because you've used up all the digits in the dividend. You can conclude that 7 goes into 3647 exactly 521 times. Alternatively, you can say that $3647/7 = 521$.**
- **If you want reassurance, check the quotient using your calculator.**

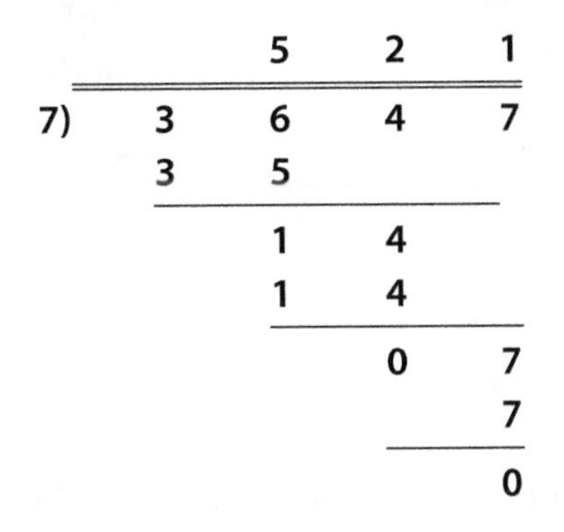

TIP *Some teachers dislike the expression "goes into." But "goes into" does a good job of describing how things work, doesn't it?*

If you don't have a calculator handy (and you wouldn't bother with longhand division if you had one, would you?), it's a good idea to check the

foregoing result by *multiplying back*, calculating the product 521×7. If you did all your work correctly, you'll get 3647. Here's the multiplication matrix:

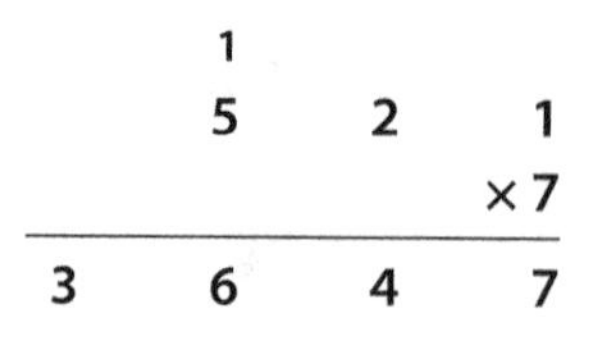

Now let's see what happens when we work through a longhand division problem and the answer doesn't come out as a whole number. We'll have to figure out what to do with the non-whole-number part of the answer, the part that "spills over" in the form of a *remainder*. In long-division problems, remainders occur most of the time, and in some cases, they get rather complicated.

PROBLEM 5-14

Use the matrix method for longhand division to calculate the quotient of 3650 divided by 7. Then figure out what to do with the quantity left over when the answer doesn't turn out as a whole number.

SOLUTION

The following steps describe the work. As you read through it all, keep track of events in the matrix below. For awhile, it'll move along exactly as the solution to Problem 5-13 did. Then things will suddenly change.

- Figure out the largest whole-number multiple of the divisor (7) that goes into the first digit of the dividend (3). The answer is none. You can't do anything here.
- Figure out the largest whole-number multiple of the divisor (7) that goes into the quantity represented by the first two digits of the dividend (36). The answer is 5.
- Write down the digit 5 in the answer row above the digit 6 in the numeral 3650. Then note that $5 \times 7 = 35$. Write down 35 below the digits 36 in the numeral 3650.
- Subtract $36 - 35$ to get 1, and bring down the 5 from the dividend to make 15.
- Determine the largest whole-number multiple of 7 that goes into 15. The answer is 2.

- **Write the digit 2 after the digit 5 in the answer row. Then note that $2 \times 7 = 14$. Write the digits 14 below the 15 that you got when you subtracted $36 - 35$ and brought down the 5.**
- **Subtract $15 - 14$ to get 1, and bring down the 0 from the dividend to make 10.**
- **Determine the largest whole-number multiple of 7 that you can fit into 10. The answer is 1.**
- **Write down the digit 1 after the digit 2 in the answer row. Then note that $1 \times 7 = 7$. Write down this digit 7 below the 10 that you got when you subtracted $15 - 14$ and brought down the 0.**
- **Subtract $10 - 7$ to get 3. Because this quantity doesn't equal 0, you know that 521 is *not* the exact answer. You have a remainder of 3.**
- **Write down the remainder after the answer 521 as "+ r 3."**

```
          5   2   1   + r 3
      ===============
7)    3   6   5   0
      3   5
      -------------
          1   5
          1   4
          ---------
              1   0
                  7
              -----
                  3
```

This remainder represents 3 parts of the divisor 7, which equals the fraction 3/7. You can, therefore, rewrite the solution as 521-3/7.

Still Struggling

With respect to the commutative law, division suffers from the same shortcoming as subtraction: It doesn't work! In general, you can't divide two numbers in opposite orders and get the same quotient both ways. If you transpose the divisor and the dividend in a division problem (when neither quantity equals 0), you get the *reciprocal* (or *multiplicative inverse*) of the result you got before. The reciprocal of a number equals 1 divided by that number. When you multiply any nonzero quantity by its reciprocal, you will get 1.

PROBLEM 5-15

Now try a longhand division problem in which the divisor has two digits. Use the matrix method for longhand division to work out 2856 divided by 37, and express the remainder as a fraction.

SOLUTION

Set up the problem as shown below. You can read the expression "37) 2856" out loud as "37 goes into 2856." It represents the ratio or fraction 2856 / 37.

	(Answer will go here)			
37)	2	8	5	6

The following steps describe the work. As you read through it all, keep track of your progress in the matrix below.

- Determine the largest whole-number multiple of the divisor (37) that goes into the first digit of the dividend (2). The answer is none, so you can't do anything.
- Determine the largest whole-number multiple of the divisor (37) that goes into the first two digits of the dividend (28). The answer is still none, so you still can't do anything.
- Determine the largest whole-number multiple of the divisor (37) that goes into the quantity represented by the first three digits of the dividend (285). You'll probably have to do several "test multiplications" (the longhand way, of course) to figure out that the answer is 7.
- Write down the digit 7 in the answer row above the digit 5 in the dividend numeral 2856. Then note that $7 \times 37 = 259$. Write down 259 below the digits 285 in the numeral 2856.
- Subtract $285 - 259$ to get 26. (You'll have to borrow in this subtraction problem; for clarity, the borrowing process is not shown here). Bring down the 6 from the dividend to make 266.
- Find the largest whole-number multiple of 37 that goes into 266. The answer is 7.
- Write down a second digit 7 after the existing digit 7 in the answer row. Then note that $7 \times 37 = 259$. Write down these digits 259 below the quantity 266 that you got when you subtracted $285 - 259$ and brought down the 6.

- Subtract 266 − 259 to get 7. This quantity indicates a remainder of 7.
- Write down the remainder after the answer 77 as "+r7."

			7	7	+r7
37)	2	8	5	6	
	2	5	9		
		2	6	6	
		2	5	9	
				7	

The remainder represents 7 parts of the divisor 37, which gives you the fraction 7/37. You can, therefore, portray the final quotient as 77-7/37.

PROBLEM 5-16

Use the matrix method for longhand division to work out 8374 divided by 109, and express the remainder as a fraction.

SOLUTION

The following steps describe the work. As you read through, keep track of your progress in the matrix below.

- Determine the largest whole-number multiple of the divisor (109) that goes into the first digit of the dividend (8). The answer is none, so you can't do anything.
- Determine the largest whole-number multiple of the divisor (109) that goes into the quantity represented by the first two digits of the dividend (83). The answer is none again, so you still can't do anything.
- Figure out the largest whole-number multiple of the divisor (109) that goes into the quantity represented by the first three digits of the dividend (837). After a few "tests," you'll discover that the answer is 7.
- Write down the digit 7 in the answer row above the digit 7 in the numeral 8374. Then note that 7 × 109 = 763. Write down 763 below the digits 837 in the numeral 8374.
- Subtract 837 − 763 to get 74. (You'll have to borrow in this subtraction problem; for clarity, the borrowing process isn't shown here). Bring down the 4 from the dividend to make 744.

- Find the largest whole-number multiple of 109 that goes into 744. The answer is 6.
- Write down a digit 6 after the digit 7 in the answer row. Then note that $6 \times 109 = 654$. Write down the digits 654 below the quantity 744 that you got when you subtracted 837 − 763 and brought down the 4.
- Subtract 744 − 654 to get 90 as the remainder.
- Write down the remainder after the answer 76 as "+r90."

			7	6	+r90
109)	8	3	7	4	
	7	6	3		
		7	4	4	
		6	5	4	
			9	0	

This remainder represents 90 parts of the divisor 109, or 90/109. You can now portray the solution as 76-90/109.

TIP *When a decimal point appears in the dividend of a longhand division problem (but not in the divisor), you can set up the matrix by "carrying the decimal point down" through the entire column where it originally appears—all the way to the bottom. You also "copy it up" into the answer row. Once you've arranged things that way, you can work through the problem exactly as you would do if the decimal point didn't exist at all. Determining the effective value of the remainder can get a little tricky, though. In the following two problems, you'll see how things work when the dividend (but not the divisor) contains a decimal point.*

PROBLEM 5-17

Use the longhand matrix method to calculate 58.4 divided by 9. Express the answer with a remainder, even if that remainder doesn't turn out as a whole number.

SOLUTION

The following steps describe the work. As you read through it all, keep track of your progress in the matrix below.

- Determine the largest whole-number multiple of the divisor (9) that goes into the first digit of the dividend (5). The answer is none, so you can't do anything.

- **Determine the largest whole-number multiple of the divisor (9) that goes into the first two digits of the dividend (58). The answer is 6.**
- **Write down the digit 6 in the answer row above the digit 8 in the dividend numeral 58.4. Then note that $6 \times 9 = 54$. Write down 54 below the digits 58 in the dividend numeral 58.4.**
- **Subtract $58 - 54$ to get 4. Bring down the decimal point and the last digit (.4) from the dividend to make 4.4. Then, for a moment, forget about the decimal point, and think of this quantity as 44.**
- **Determine the largest whole-number multiple of 9 that goes into 44. The answer is 4.**
- **Write down a digit 4 after the digit 6 in the answer row, right above the 4 in the dividend numeral, and right after the decimal point. Then note that $4 \times 9 = 36$. Write down these digits below the numeral 4.4, but put a decimal point between them so that you actually write 3.6.**
- **Subtract $4.4 - 3.6$ to get 0.8 as the remainder.**
- **Write down the remainder after the answer 6.4 as "+ r 0.8."**

```
             6 . 4   + r 0.8
      ==============
9)    5      8 . 4
      5      4 .
      --------------
             4 . 4
             3 . 6
             -------
             0 . 8
```

Still Struggling

The above-derived remainder represents 0.8 parts of the divisor, which equals 9 in this case. You can regard it as 0.8/9, but that's not a good way to denote a fraction. There's an easy way around this glitch, however. You can multiply the numerator and denominator both by 10 to get 8/90, which is a perfectly decent fraction. The quotient equals $6.4 + 8/90$. If you convert the fraction 8/90 to decimal form, you get $0.088888\cdots$ (endlessly repeating). Now you can add everything up and see that the quotient equals $6.4 + 0.088888\cdots = 6.488888\cdots$.

PROBLEM 5-18

Use the longhand matrix method to calculate 78.9 divided by 15. Then figure out what to do with the remainder.

SOLUTION

The following steps describe the work. As you read through the description, keep track of your progress in the matrix below.

- Determine the largest whole-number multiple of the divisor (15) that goes into the first digit of the dividend (7). The answer is none, so you can't do anything.
- Determine the largest whole-number multiple of the divisor (15) that goes into the first two digits of the dividend (78). The answer is 5.
- Write down the digit 5 in the answer row above the digit 8 in the dividend numeral 78.9. Then note that $5 \times 15 = 75$. Write down 75 below the digits 78 in the dividend numeral 78.9.
- Subtract $78 - 75$ to get 3. Bring down the decimal point and the last digit (.9) from the dividend to make 3.9. Then, for a moment, forget about the decimal point, and think of this quantity as 39.
- Find the largest whole-number multiple of 15 that goes into 39. The answer is 2.
- Write down a digit 2 after the digit 5 in the answer row, right above the 9 in the dividend numeral, and right after the decimal point. Then note that $2 \times 15 = 30$. Write down these digits below the numeral 3.9, but put a decimal point between them so you actually write 3.0.
- Subtract $3.9 - 3.0$ to get 0.9 as the remainder.
- Write down the remainder after the answer 5.2 as "+r 0.9."

```
            5 . 2   +r0.9
      ===============
15)   7     8 . 9
      7     5 .
      ---------------
            3 . 9
            3 . 0
            -------
            0 . 9
```

Once again, your remainder doesn't work out as a whole number. It represents 0.9 parts of the divisor 15, which you can write as 0.9/15. When you multiply the numerator and denominator both by 10 to get 9/150, you get a fraction in the proper form. You can now say that the quotient equals 5.2 + 9/150. When you convert 9/150 to a decimal quantity, you get 0.06. Now you can add everything up and see that the quotient equals 5.2 + 0.06 = 5.26.

TIP *So far, you haven't had to worry about division problems in which a decimal point appears in the divisor (the equivalent of the denominator in a fraction). But you should know how to deal with a situation like that. When you encounter a decimal point in the divisor of a longhand division problem, move the decimal point all the way over to the far right-hand end of the divisor so that it "disappears." Count the number of places that you had to move the point to get rid of it in the divisor. Then move the point in the dividend to the right by the same number of places. Let's try two problems of this sort.*

PROBLEM 5-19

Use the matrix method to work out 9.65 divided by 4.2. Then figure out what to do with the remainder.

SOLUTION

In this case, you have to move the decimal point to the right by one place in the divisor, 4.2, to get 42, so you'll want to move the decimal point to the right by one place in the dividend, getting 96.5. Now your starting array looks like this:

	(Answer will go here)			
42)	9	6	.	5

The following steps describe the work. As you read through it all, keep track of your progress in the matrix below.

- Determine the largest whole-number multiple of the divisor (42) that goes into the first digit of the dividend (9). The answer is none, so you can't do anything.
- Determine the largest whole-number multiple of the divisor (42) that goes into the first two digits of the dividend (96). The answer is 2.
- Write down the digit 2 in the answer row above the digit 6 in the dividend numeral 96.5. Then note that $2 \times 42 = 84$. Write down 84 below the digits 96 in the dividend.

- Subtract 96 − 84 to get 12. Bring down the decimal point and the last digit (.5) from the dividend to make 12.5. Then, for a moment, forget about the decimal point, and think of this quantity as 125.
- Find the largest whole-number multiple of 42 that goes into 125. The answer is 2.
- Write down a digit 2 after the digit 5 in the answer row, right above the 5 in the dividend numeral, and right after the decimal point. Then note that $2 \times 42 = 84$. Write down these digits below the numeral 12.5, but put a decimal point between them so you actually write 8.4.
- Subtract 12.5 − 8.4 to get 4.1 as the remainder.
- Write down the remainder after the answer 2.2 as "+r4.1".

```
          2 . 2   +r4.1
     ==========
42)  9    6 . 5
     8    4 .
     ----------
     1    2 . 5
          8 . 4
          -----
          4 . 1
```

The remainder represents 4.1 parts of the divisor 42, which you can write as 4.1/42. If you multiply the numerator and denominator both by 10, you get the legitimate fraction 41/420. You can now say that the quotient equals 2.2 + 41/420. When you convert 41/420 to decimal form, you get the following expression:

$$0.0976190476190476190476190476 1 \cdots$$

The digit sequence "761904" keeps on repeating forever, starting after the initial 0.09. That repetition produces a numerical monster! If you don't mind approximating, you can round it off to a few decimal places, say to 0.097619. Now you can add everything up and see that the quotient equals approximately 2.2 + 0.097619 = 2.297619.

TIP *You've manually calculated that 96.5/42 equals approximately 2.297619, so you know that the original quotient, 9.65/4.2, also equals approximately 2.297619.*

PROBLEM 5-20

Use the matrix method to work out 18.27 divided by 0.34. Then figure out what to do with the remainder.

SOLUTION

In this case, you must move the decimal point to the right by two places in the divisor, 0.34, to get 34. You must, therefore, move the decimal point to the right by two places in the dividend as well, getting 1827 and the following set-up array:

	(Answer will go here)			
34)	1	8	2	7

You have an easy problem, now! As you read through the following description, keep track of the steps in the matrix that appears afterward.

- Determine the largest whole-number multiple of the divisor (34) that goes into the first digit of the dividend (1). The answer is none, so you can't do anything.
- Determine the largest whole-number multiple of the divisor (34) that goes into the first two digits of the dividend (18). The answer is still none, so you still can't do anything.
- Determine the largest whole-number multiple of the divisor (34) that goes into the quantity represented by the first three digits of the dividend (182). You'll probably have to do two or three "test multiplications" to figure out that the answer is 5.
- Write down the digit 5 in the answer row above the digit 2 in the dividend numeral 1827. Then note that $5 \times 34 = 170$. Write down 170 below the digits 182 in the numeral 1827.
- Subtract 182 − 170 to get 12. Bring down the 7 from the dividend to make 127.
- Find the largest whole-number multiple of 34 that goes into 127. The answer is 3.
- Write down a digit 3 after the digit 5 in the answer row. Then note that $3 \times 34 = 102$. Write down the digits 102 below the quantity 127 that you got when you subtracted 182 − 170 and brought down the 7.
- Subtract 127 − 102 to get 25. This quantity indicates a remainder of 25.

- Write down the remainder after the answer 53 as "+r 25."
- You've calculated the quotient of 1827/34 as 53-25/34.

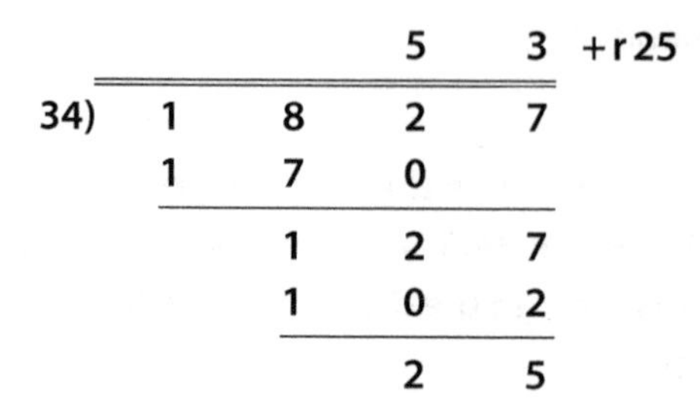

If you prefer to get a decimal expression for your answer, you can convert 25/34 to decimal form, but you'll generate a 16-digit repeating sequence! You might rather round things off to get, say, 53.73529.

TIP *You've manually calculated that 1827/34 equals about 53.73529, so you know that the original quotient, 18.27/0.34, also equals about 53.73529.*

QUIZ

Refer to the text in this chapter if necessary. A good score is eight correct. Answers are in the back of the book. (Don't use your calculator to see if the results stated here are actually correct or not! Rely on your pencil, paper, and brain alone.)

1. Where have we made borrowing errors in the following longhand subtraction problem?

$$\begin{array}{ccccccc} & 9 & 9 & 9 & & 9 & 9 \\ 1 & \cancel{0} & \cancel{0} & \cancel{0} & . & \cancel{0} & \cancel{0} \\ & -6 & 5 & 9 & . & 7 & 7 \\ \hline 1 & 3 & 4 & 0 & . & 2 & 2 \end{array}$$

A. In the leftmost column only.
B. In the leftmost and rightmost columns.
C. In all the columns except the leftmost one.
D. In all the columns except the leftmost and rightmost ones.

2. Where have we made borrowing errors in the following longhand subtraction problem?

$$\begin{array}{ccccccc} & 8 & 8 & 8 & & 8 & 8 \\ 1 & \cancel{9} & \cancel{9} & \cancel{9} & . & \cancel{9} & \cancel{9} \\ & -2 & 3 & 5 & . & 4 & 7 \\ \hline 1 & 6 & 5 & 3 & . & 4 & 1 \end{array}$$

A. In the leftmost column only.
B. In the leftmost and rightmost columns.
C. In all the columns except the leftmost one.
D. In all the columns except the leftmost and rightmost ones.

3. What mistake, if any, have we made in the following longhand addition problem?

$$\begin{array}{ccccc} & 9 & 8 & 7 & 6 \\ & +6 & 7 & 8 & 9 \\ \hline 1 & 5 & 5 & 5 & 5 \end{array}$$

A. We forgot to borrow digits where necessary.
B. We forgot to carry digits where necessary.
C. We forgot to carry and borrow digits where necessary.
D. We haven't made any mistake.

4. **What mistake or mistakes have we committed in the following longhand multiplication problem?**

	2		
	2	3	1
		×2	8
1			
1	8	7	8
+4	6	2	0
6	4	9	8

A. We've done one of the single-digit multiplication steps wrong.
B. We've done one of the single-digit addition steps wrong.
C. We've forgotten to carry a digit where necessary.
D. All of the above

5. **If we transpose the order of the quantities in a difference (subtraction), the result**

A. stays the same as it was before.
B. turns into the negative of what it was before.
C. turns into the reciprocal of what it was before.
D. becomes equal to 1.

6. **What mistake or formatting inconsistency, if any, have we committed in the following longhand multiplication problem?**

			2	
		2	4	
		1	0	8
		×3	0	6
	1			
		8	4	8
	+0	0	0	0
+3	2	4	0	0
3	3	2	4	8

A. We've done one of the single-digit multiplication steps wrong, and also carried a digit improperly.
B. We've done one of the single-digit addition steps wrong.
C. We've aligned one of the rows improperly, writing it down one decimal place too far to the left.
D. We haven't committed any mistake or formatting inconsistency.

7. **What mistake or formatting inconsistency, if any, have we committed in working through the following longhand addition problem?**

		1	2	.	0	2
	+2	0	0	.	2	4
+1	2	7	3	.	6	3
1	4	8	5	.	8	9

A. We've forgotten to borrow digits where necessary.
B. We've forgotten to carry digits where necessary.
C. We've forgotten to carry and borrow digits where necessary.
D. We haven't committed any mistake or formatting inconsistency.

8. What mistake or formatting inconsistency, if any, have we committed in working through the following longhand division problem?

		6	7	1
6)	4	0	2	6
	3	6		
		4	2	
		4	2	
			0	6
				6
				0

A. We've made an intermediate multiplication error.
B. We've forgotten to carry a digit when necessary.
C. We've improperly borrowed digits when we shouldn't have.
D. We haven't committed any mistake or formatting inconsistency.

9. If we transpose the dividend and divisor in a quotient, assuming that neither of them equals 0, the result

A. stays the same as it was before.
B. turns into the negative of what it was before.
C. turns into the reciprocal of what it was before.
D. becomes equal to 1.

10. What mistake or formatting inconsistency, if any, have we committed in working through the following longhand division problem?

		4	.	1	+r1.4
17)	7	5	.	1	
	7	2	.		
		3	.	1	
		1	.	7	
		1	.	4	

A. We've made an intermediate multiplication error.
B. We've forgotten to carry a digit when necessary.
C. We've improperly borrowed digits when we shouldn't have.
D. We haven't committed any mistake or formatting inconsistency.

chapter 6

Objects

In everyday life, you'll occasionally need to figure out the size, expanse, or volume of an object, region, or space. For example, you might want to know how much carpet you'll need to cover your living room's floor, or how many cubic inches of "stuff" you can pack into your foot locker for transport.

CHAPTER OBJECTIVES

In this chapter, you will

- Identify basic shapes that can exist on flat surfaces.
- Learn to calculate the perimeters and areas of right triangles, squares, and rectangles.
- Learn to calculate the circumferences and areas of circles.
- Identify basic solids that can exist in physical space.
- Learn to calculate the surface areas and volumes of cubes, boxes, and spheres.

Perimeter and Circumference

In the "real world," you'll encounter objects or regions that lie entirely on a single flat surface called a *plane*. Here are the most common examples.

- A *right triangle* has three straight sides whose lengths can differ, three *vertices* (points where sides come together), and three angles, one of which is a *right angle* (meaning that it measures 90°, or 1/4 of a complete circle).
- A *square* has four straight sides, all of the same length, four vertices, and four angles, all of the same measure.
- A *rectangle* has four straight sides, opposite pairs of which have the same length, four vertices, and four angles, all of the same measure.
- A *circle* comprises a perfectly round, closed curve, representing all points at a certain fixed distance from a defined center.

The *perimeter* of a straight-sided figure equals the sum of the lengths of all the sides. The *circumference* of a circle equals the distance going once around the curve. For any object with straight sides (technically called a *polygon*), we can calculate the perimeter easily. Calculating the circumference of a circle poses a slightly more complicated problem, but it's easy once you know the trick.

PROBLEM 6-1

How can we calculate the perimeter of a right triangle? If we find a right triangle whose sides measure 3, 4, and 5 meters, what's the perimeter?

SOLUTION

Figure 6-1 shows a right triangle where the sides have lengths a, b, and c, and the 90° angle lies opposite the longest side. The perimeter equals the sum of the lengths of the sides, $a + b + c$.

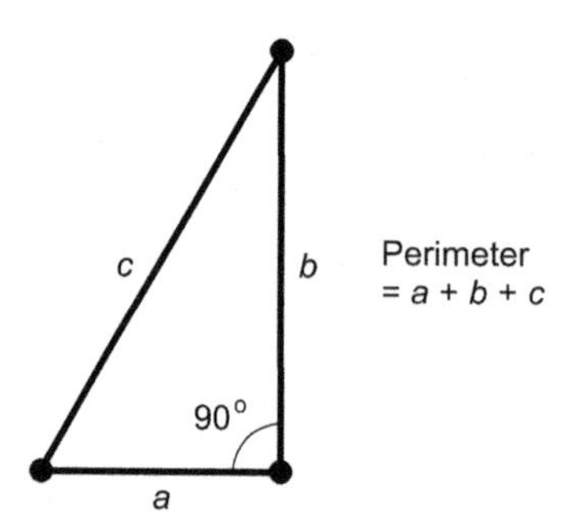

FIGURE 6-1 • Perimeter of a right triangle. Illustration for Problem 6-1 and its solution.

Suppose that we come across a right triangle where $a = 5$ meters, $b = 4$ meters, and $c = 3$ meters, for example. In that case, the whole triangle has a perimeter of $3 + 4 + 5 = 12$ meters.

Here's a Factoid!

In a right triangle, the lengths of the sides have a certain relationship. If we multiply each individual side's length by itself, we'll always find that

$$aa + bb = cc$$

when c represents the length of the longest side, which mathematicians call the *hypotenuse*. When we multiply something by itself, we *square* it, and we can represent that fact by writing a superscript 2 after it. A mathematician would rewrite the above equation as

$$a^2 + b^2 = c^2$$

This equation holds true no matter how "tall and thin" or "short and fat" the right triangle gets. It has been known as a *theorem* (proven fact in mathematics) for thousands of years. According to legend, the person who originally proved it was a Greek named *Pythagoras*. To this day, people refer to the *Theorem of Pythagoras* (or the *Pythagorean Theorem*) when talking about right triangles.

PROBLEM 6-2

How can we calculate the perimeter of a square? If we find a square whose sides each measure exactly 1.2 inches long, what's the perimeter?

SOLUTION

Figure 6-2 shows a generic square where the sides have lengths a, b, c, and d, and the interior angles have measures w, x, y, and z. In any square with sides and angles as named in Fig. 6-2, we know for certain that $a = b = c = d$, and also that $w = x = y = z$.

In a square, the sides can have any length from a fraction of a millimeter to millions of kilometers provided that they're all equally long, but all four angles measure 90° no matter what.

The perimeter of the square in Fig. 6-2 equals the sum of the lengths of the sides, $a + b + c + d$. Because all four sides have the same length, the perimeter equals four times the length of any one side.

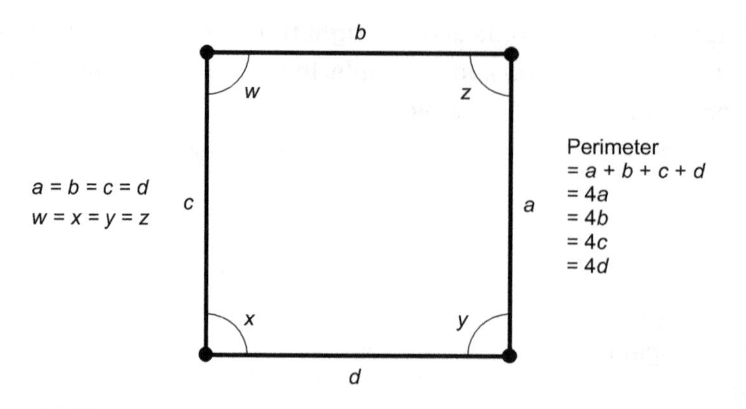

FIGURE 6-2 • Perimeter of a square. Illustration for Problem 6-2 and its solution.

If the sides of a square all measure 1.2 inches, for example, then the entire figure has a perimeter of $1.2 + 1.2 + 1.2 + 1.2 = 4 \times 1.2 = 4.8$ inches.

PROBLEM 6-3

How can we calculate the perimeter of a rectangle? If we find a rectangle with one pair of opposite sides measuring exactly 4 feet each and the other pair of opposite sides measuring exactly 6 feet each, what's the perimeter of the whole figure?

SOLUTION

Figure 6-3 shows a generic rectangle where the sides have lengths a, b, c, and d, and the interior angles have measures w, x, y, and z. In any rectangle with sides and angles as named in Fig. 6-3, we can have complete faith that $a = c$, $b = d$, and $w = x = y = z$.

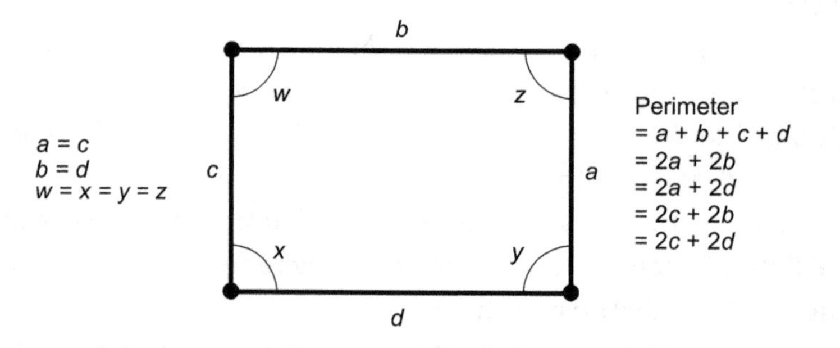

FIGURE 6-3 • Perimeter of a rectangle. Illustration for Problem 6-3 and its solution.

In a rectangle, the sides can have any lengths, provided that opposite pairs of sides are equally long. (Adjacent sides can differ in length, as we can see in Fig. 6-3.) As with the square, all four angles measure 90° no matter what.

The perimeter of the rectangle in Fig. 6-3 equals the sum of the lengths of the sides, $a + b + c + d$. Because opposite pairs of sides have equal length, we can calculate the perimeter by doubling the length of one side, and then adding twice the length of the adjacent side.

For example, if we find a rectangle where one pair of opposite sides measures 4 feet each and the other pair of opposite sides measures 6 feet each, then the perimeter equals $4 + 4 + 6 + 6 = (2 \times 4) + (2 \times 6) = 8 + 12 = 20$ feet.

TIP *A square is a specific type of rectangle, in which all four sides have the same length. All squares are rectangles, but not all rectangles are squares.*

PROBLEM 6-4

How can we calculate the circumference of a circle? If we find a circle whose radius equals precisely 5 inches, what's the circumference?

SOLUTION

Figure 6-4 illustrates a circle with a radius that we call r, and a circumference that we call b. For any circle that lies on a flat surface, no matter how large or small, we'll find that b equals a certain multiple of r. That multiple always turns out as twice the value of a well-known mathematical constant called *pi* (symbolized with the lowercase Greek letter π). For most practical purposes, we can say that $\pi = 3.1416$.

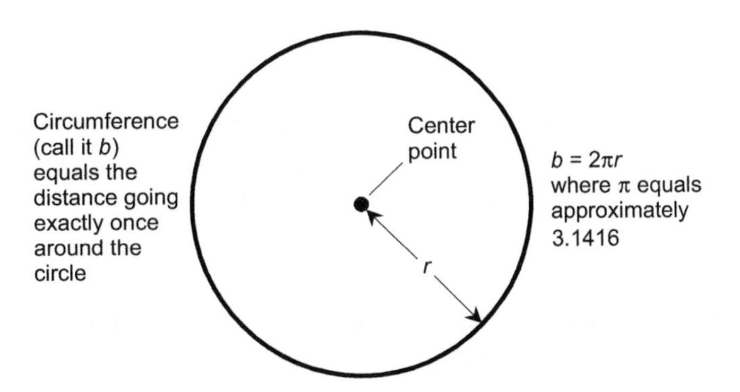

FIGURE 6-4 • Circumference of a circle. Illustration for Problem 6-4 and its solution.

To calculate the circumference of a circle whose radius equals exactly 5 inches, we must multiply $2 \times 3.1416 \times 5$. That arithmetic gives us a product of 31.416 inches. That's what we'd observe if we took a flexible tape measure to determine the distance going around the circle along its periphery, from any particular point back to that same point.

TIP *If you have a scientific calculator that can deal with the so-called* inverse trigonometric functions, *you can get it to display* π *with great accuracy. Set the calculator to work in radians (not in degrees). Then find the* inverse cosine *(inv cos or* $\cos^{-1}$*) of* -1*. Your display should show 3.14159 and then some more digits. A personal computer's calculator can display* π *out to more decimal places than you'll ever need!*

Still Struggling

The constant π is one of those enigmatic quantities that mathematicians call an irrational number or nonterminating, nonrepeating decimal. We only approximate its value in the "real world." After the decimal point, the sequence of digits goes on forever, but it never follows any pattern (at least, no pattern that anybody has found yet). We can't portray π as a fraction or ratio, although in ancient times, some mathematicians thought that it was equal to 22/7.

Interior Area

We can calculate the *interior area*, or simply the *area*, of a polygon using a formula tailored to that particular type of figure. We can calculate the area of a circle with a formula that applies to all circles. Let's look at the object types from the previous section, and learn how to determine their areas.

TIP ***Each one of the following formulas will give you the correct results if, but only if, the applicable figure and its interior lie entirely on a flat surface. Formulas exist for figures that lie on "curved" or "warped" surfaces, such as cylinders, cones, and spheres; but they're complicated, and they go far beyond the scope of this book!***

PROBLEM 6-5

How can we calculate the interior area of a right triangle? If we find a right triangle whose sides measure 3, 4, and 5 inches, what's the area?

SOLUTION

Figure 6-5 shows a right triangle where the sides have lengths a, b, and c, and the 90° angle lies opposite the longest side. It's the same object as the one shown in Fig. 6-1, except that we've shaded the interior to indicate that we want to calculate the area, not the perimeter.

For a triangle, the area equals the product of the base length and the height divided by 2, where we define the height going perpendicular (at a right angle) to the base. A right triangle gives us an easy job because two of the three sides run perpendicular to each other. In the situation of Fig. 6-5, a represents the base length and b represents the height, so the area equals $ab/2$.

If $a = 3$ inches and $b = 4$ inches, the right triangle has an area of $3 \times 4/2 = 12/2 = 6$ square inches. If we turn the triangle on its side and make b represent the base length and a represent the height, we'll get the same result for the interior area. (Try it and see.)

TIP *When you want to quantify the interior area of any geometric figure, you must use* square units *(which some scientists call* units squared*). For example, if you have straight-line measures in inches, you'll want to specify the area in square inches or inches squared. If you have straight-line measures given in meters, you'll want to calculate the area in square meters or meters squared.*

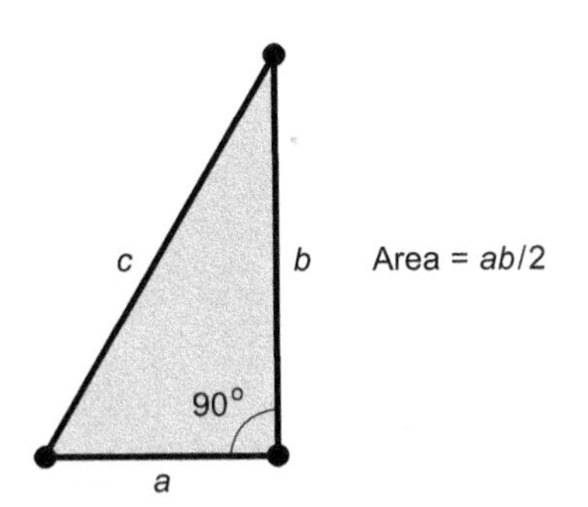

FIGURE 6-5 • Interior area of a right triangle. Illustration for Problem 6-5 and its solution.

Still Struggling

If your straight-line measures aren't both given in the same units, you should convert both quantities to the same unit before trying to calculate the area of the object in question. For example, you might discover that the triangle in Fig. 6-5 has a base length of 3 inches and a height of 1/3 of a foot (which happens to equal 4 inches); you can convert both the base length and the height to inches, and calculate the area in square inches.

PROBLEM 6-6

How can we calculate the interior area of a square? If we find a square whose sides each measure exactly 1.2 miles long, what's the area?

SOLUTION

Figure 6-6 shows a square where the sides measure *a*, *b*, *c*, and *d*, and the angles measure *w*, *x*, *y*, and *z*. It's the same square as the one in Fig. 6-2, except that we've shaded the interior to emphasize that we want to find the area rather than the perimeter.

The area enclosed by a square equals the product of the lengths of any two adjacent sides. In a square, all four sides have the same length, so the area equals the square of the length of any one side (the length multiplied by itself).

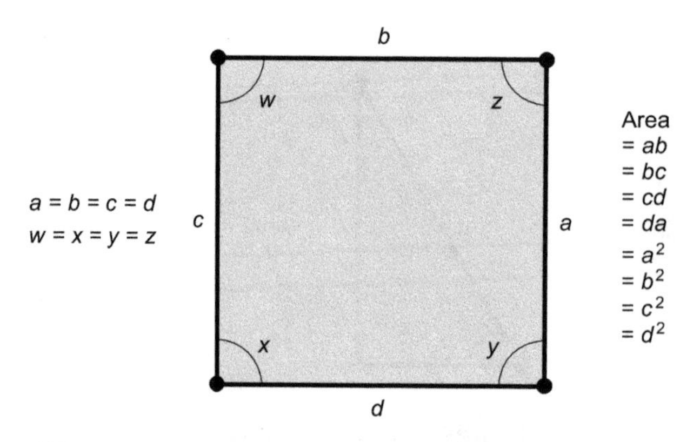

FIGURE 6-6 • Interior area of a square. Illustration for Problem 6-6 and its solution.

If the sides of a square, such as the one shown in Fig. 6-6, all measure 1.2 miles, we can calculate the area as ab, bc, cd, da, a^2, b^2, c^2, or d^2 (eight different schemes, with the same result every time). We always get $1.2 \times 1.2 = 1.2^2 = 1.44$ square miles.

PROBLEM 6-7

How can we calculate the interior area of a rectangle? If we find a rectangle with one pair of opposite sides measuring exactly 5 centimeters each and the other pair of opposite sides measuring exactly 7 centimeters each, what's the area of the figure?

SOLUTION

Figure 6-7 shows a rectangle where the sides have lengths a, b, c, and d, and the interior angles have measures w, x, y, and z. It's the same object that we saw in Fig. 6-3, except that we've shaded the interior to emphasize the fact that we seek the area enclosed by the rectangle, and not the length of its perimeter.

The area of a rectangle equals the product of the lengths of any two adjacent sides. Because opposite pairs of sides have equal length, we can calculate the area in Fig. 6-7 as ab, bc, cd, or da (four different processes, with the same result every time).

If we encounter a rectangle where one pair of opposite sides measures 5 centimeters each and the other pair of opposite sides measures 7 centimeters each, then we can calculate the interior area as $5 \times 7 = 35$ square centimeters.

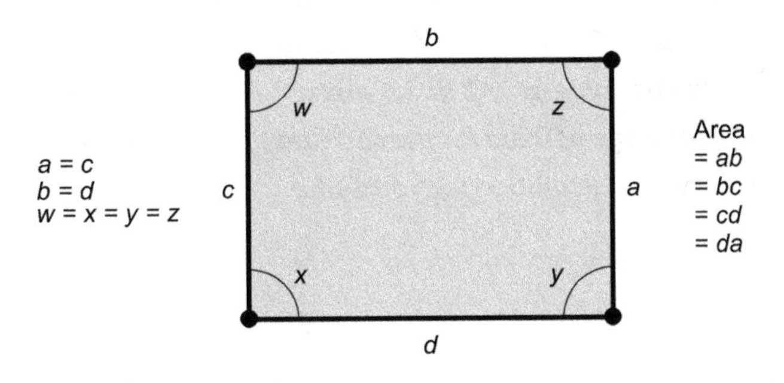

FIGURE 6-7 • Interior area of a rectangle. Illustration for Problem 6-7 and its solution.

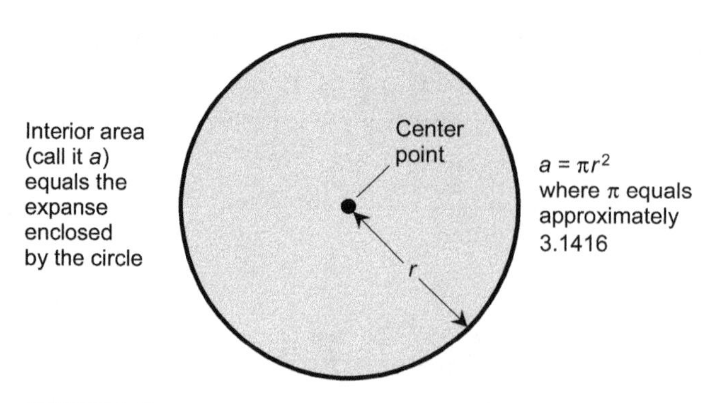

FIGURE 6-8 • Interior area of a circle. Illustration for Problem 6-8 and its solution.

PROBLEM 6-8

How can we calculate the area enclosed by a circle? If we find a circle whose radius equals precisely 7 meters, what's the circumference, rounded off to the nearest square meter?

SOLUTION

Figure 6-8 illustrates a circle with a radius that we call *r*, and an area that we call *a*. It's the same object that we saw in Fig. 6-4, except that we've shaded it to emphasize the fact that we want to find the interior area, rather than the circumference.

For any circle that lies on a flat surface, its area *a* equals π times the square of the radius *r*. To calculate the area of a circle whose radius equals exactly 7 meters, we can multiply $3.1416 \times 7 \times 7$ to get 154 square meters, rounded off to the nearest square meter.

TIP ***Remember that in a circle, the radius equals half the diameter! If the foregoing problem had asked for the area of a circle whose diameter is 7 meters, we'd have to set the radius value at 7/2 or 3.5 meters. Then the resulting area would turn out only 1/4 as large as it did for the circle having the radius 7 meters. (If you don't believe it, try the calculation for yourself.)***

Surface Area

In three dimensions, we encounter a wide variety of objects. Here are the most common examples. If we consider the object's surface only and not the interior, we call it a *shell*. If we think about the surface along with the interior, we call the whole thing a *solid*.

- A *cube* has six identical square faces, 12 edges, and eight vertices (points where adjacent edges converge). The angle between any two adjacent edges is a right angle (it measures 90°). All of the edges are equally long.
- A *box* has six rectangular faces, 12 edges, and eight vertices. The angle between any two adjacent edges is a right angle. The edges don't all have to be the same length.
- A *sphere* constitutes a perfectly round object, comprising all the points at, or less than, a certain distance from a center point. Some people use the term *ball* instead of sphere, although this usage can get confusing because in sports, some "balls" aren't spheres. (Think of an American football.)

The *surface area* of a cube or box equals the sum of the interior areas of all its faces. The surface area of a sphere equals the area that we'd observe if we could flatten out its shell without stretching it. For any solid object with straight edges and flat faces (technically called a *polyhedron*), we can calculate the surface area by adding up the areas of all the faces. Calculating the surface of a sphere involves a trick, similar to what we do when we find the circumference of a circle.

PROBLEM 6-9

How can we calculate the surface area of a cube? If we find a cube whose edges each measure exactly 1.2 inches long, what's the surface area of the entire object?

SOLUTION

Figure 6-9 shows the shell of a cube with height of measure *a*, width of measure *b*, and depth of measure *c*. At the vertex points, the angles

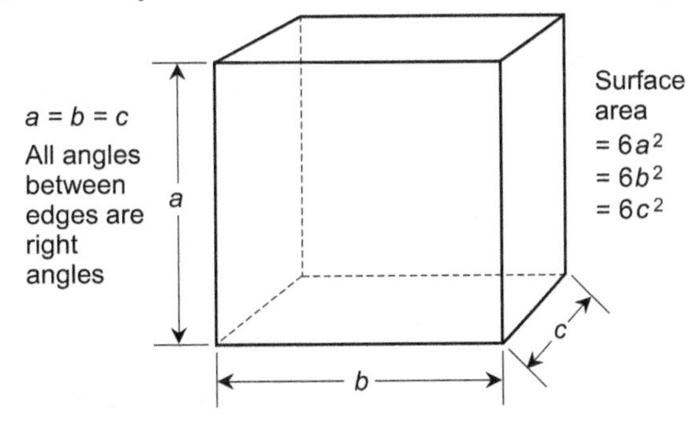

FIGURE 6-9 • Surface area of a cube. Illustration for Problem 6-9 and its solution.

between pairs of edges are always right angles. In any cube with edges as named in Fig. 6-9, we know that $a = b = c$.

The surface area of a cube equals six times the area of any given face (the object has six identical faces), that is, $6a^2$, $6b^2$, or $6c^2$, as shown in Fig. 6-9. Every face forms a perfect square, so the area of any given face equals the square of the length of any edge.

If the edges of the cube all measure 1.2 inches, then the object has a surface area of $6 \times 1.2^2 = 6 \times 1.44 = 8.64$ square inches.

PROBLEM 6-10

How can we calculate the surface area of a box? If we find a box that's exactly 4 meters high, 6 meters wide, and 3 meters deep, what's the surface area of the whole object?

SOLUTION

Figure 6-10 shows the shell of a box with height a, width b, and depth c. As with the cube, the angles between pairs of edges at the vertex points are always right angles. Unlike the cube, however, the height, width, and depth can all differ.

The surface area of a box with dimensions as labeled in Fig. 6-10 equals twice the sum of the three pairwise products of the edges, $2(ab + bc + ac)$, or as the sum of twice the pairwise products of the edges, $2ab + 2bc + 2ac$. Every face forms a rectangle, so the area of any given face equals the product of the lengths of any two adjacent edges.

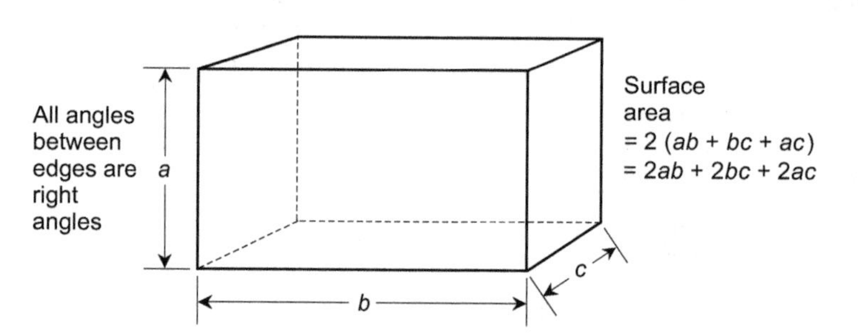

FIGURE 6-10 • Surface area of a box. Illustration for Problem 6-10 and its solution.

If we find a box precisely 4 meters high by 6 meters wide by 3 meters deep, then the surface area of the entire object works out as follows:

$$\begin{aligned}\text{Surface area} &= [2 \times (4 \times 6)] + [2 \times (6 \times 3)] + [2 \times (4 \times 3)] \\ &= (2 \times 24) + (2 \times 18) + (2 \times 12) \\ &= 48 + 36 + 24 \\ &= 108 \text{ square meters}\end{aligned}$$

TIP *A cube constitutes a special sort of box, in which all of the edges have identical lengths. All cubes are boxes, but not all boxes are cubes.*

PROBLEM 6-11

How can we calculate the surface area of a sphere? If we find a sphere whose radius equals precisely 7 meters, what's the surface area of its shell, rounded off to the nearest square meter?

SOLUTION

Figure 6-11 illustrates the shell of a sphere with radius *r* and surface area *a*. The shell's surface area equals 4π times the square of the radius. To calculate

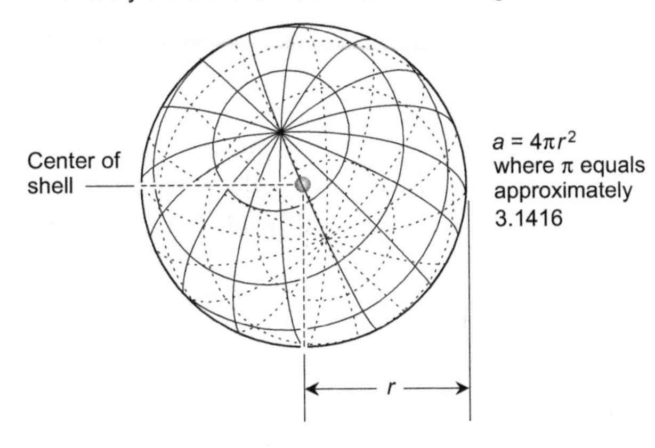

FIGURE 6-11 • Surface area of a sphere. Illustration for Problem 6-11 and its solution.

the surface area of a sphere whose radius equals exactly 7 meters, we can multiply $4 \times 3.1416 \times 7 \times 7$ to get 616 square meters, rounded off to the nearest square meter.

TIP *Remember that, in a sphere, the radius equals half the diameter, just as with a circle. If the foregoing problem had asked for the surface area of a sphere with a diameter of 7 meters, we'd have a radius of 3.5 meters. Then the surface area would turn out 1/4 as large as it did for the sphere of radius 7 meters. (If you don't believe it, try the calculation for yourself, just as you might have done with the circle.)*

Still Struggling

Surface areas always come out in square units, just as interior areas do. That's because we always multiply two (but never more than two) different straight-line measures by one another, or else multiply a single straight-line measure by itself. We call each straight-line-measurable quantity a *dimension*. Straight lines and curves are one-dimensional when we think of them in terms of distance or displacement. Flat or warped surfaces are two-dimensional when we think of them in terms of area or expanse.

Volume

The *volume* of a solid object quantifies the amount of space, in three dimensions, that it displaces. We can calculate the volume of a cube or box easily. The sphere presents a tougher problem, but a simple formula allows us to work it out with a calculator.

PROBLEM 6-12

How can we calculate the volume of a cube? If we find a cube whose edges each measure exactly 1.2 inches long, what's the volume?

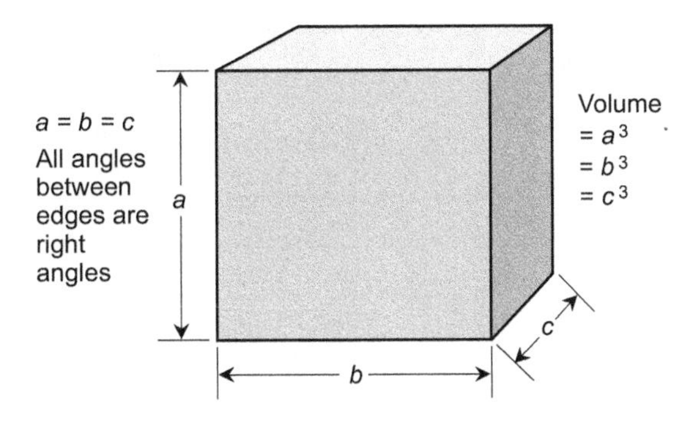

FIGURE 6-12 • Volume of a cube. Illustration for Problem 6-12 and its solution.

SOLUTION

Figure 6-12 shows a cube where the edges measure a, b, and c. It's the same object as the one in Fig. 6-9, except that we've made it appear opaque to emphasize the fact that we want to find the volume rather than the surface area.

The volume of a cube equals the product of the height, the width, and the depth. In a cube, all of the edges have the same length, so the volume equals the cube (third power) of the length of any one edge—that is, the length multiplied by itself, and then multiplied by itself again. We denote the cube of a quantity by writing a superscript 3 after it.

If the height a, the width b, and the depth c of a cube all measure 1.2 inches, we can calculate the volume as a^3, b^3, or c^3. We always get $1.2 \times 1.2 \times 1.2 = 1.2^3 = 1.728$ cubic inches.

TIP *When you want to quantify a volume, you must use* cubic units *(which some scientists call* units cubed*). For example, if you have straight-line measures in inches, you'll usually specify the volume in cubic inches or inches cubed. If you have straight-line measures given in meters, you'd calculate the volume in cubic meters or meters cubed.*

PROBLEM 6-13

How can we calculate the volume of a box? If we find a box that's exactly 4 meters high, 6 meters wide, and 3 meters deep, what's the volume of the whole object?

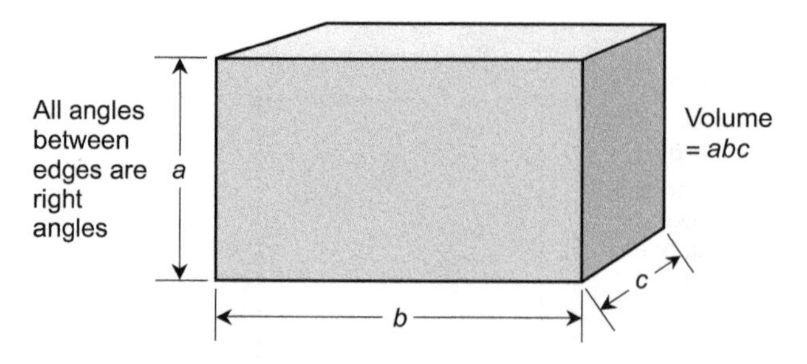

FIGURE 6-13 • Volume of a box. Illustration for Problem 6-13 and its solution.

SOLUTION

Figure 6-13 shows a solid box with height *a*, width *b*, and depth *c*. It's the same object as the one in Fig. 6-10, except we've made it look opaque to indicate that we're working with the volume of the whole solid, not the surface area of the shell.

The volume of a box with dimensions as labeled in Fig. 6-13 equals the product of the lengths of the edges defining the height, width, and depth. In this case, that's simply *abc*.

If we find a box whose dimensions measure exactly 4 meters high by 6 meters wide by 3 meters deep, then the volume of the solid is $4 \times 6 \times 3 = 72$ cubic meters.

Still Struggling

If your straight-line measures aren't all quantified in the same units, then you must convert all quantities to the same unit before trying to calculate the volume of an object. For example, you might find that the box in Fig. 6-13 has a height of 5 feet, a width of 8 feet, and a depth of 48 inches (which is 4 feet). You can convert all straight-line dimensions to feet, and calculate the volume in cubic feet. Alternatively, you can convert all straight-line dimensions to inches, and calculate the volume in cubic inches. You might get a surprise if you try both methods! A foot equals 12 inches, so a cubic foot equals $12 \times 12 \times 12 = 1728$ cubic inches. The volume of any object expressed in cubic inches will, therefore, work out as 1728 times the volume of the same object expressed in cubic feet.

PROBLEM 6-14

How can we calculate the volume of a sphere? If we find a sphere whose radius equals precisely 7 meters, what's its volume, rounded off to the nearest cubic meter?

SOLUTION

Figure 6-14 illustrates a sphere with a radius that we call r and a volume that we call v. It's the same sphere that we saw in Fig. 6-11, except that we've made it appear opaque to indicate that we seek the volume, not the surface area. The volume of any sphere equals $4\pi/3$ times the cube of its radius. To find the volume of a sphere whose radius is exactly 7 meters, we multiply $4 \times 3.1416 \times 7 \times 7 \times 7$ and then divide by 3. That little arithmetic exercise gives us 1437 cubic meters, rounded off to the nearest cubic meter.

TIP *If the preceding problem had asked for the volume of a sphere with a diameter of 7 meters, we'd have a radius of 3.5 meters. Then the volume would turn out to be only 1/8 as large as it did for the sphere with a radius of 7 meters. When we cut the radius or the diameter of a sphere in half, we reduce the volume to $(1/2) \times (1/2) \times (1/2)$, or 1/8, of its previous value. That's why, for example, hail stones measuring 2 inches in diameter do so much more damage than hail stones measuring 1 inch in diameter!*

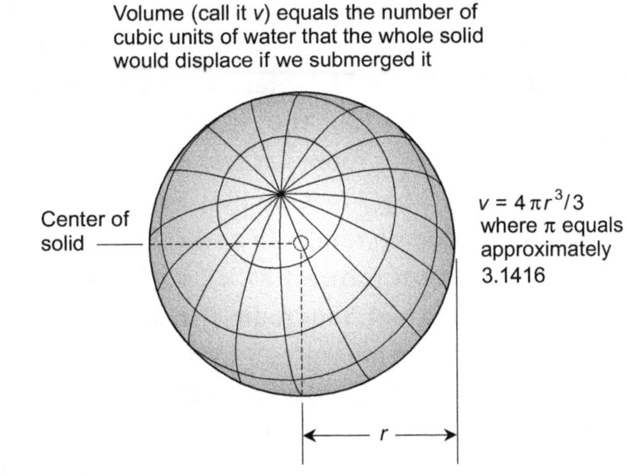

FIGURE 6-14 • Volume of a sphere. Illustration for Problem 6-14 and its solution.

Combining Areas and Volumes

You can add the areas of two regions together, and the total area will equal the sum of the areas. You can do the same thing with volumes. You can subtract one area from another, or one volume from another, and—you guessed it—the net area or volume will equal the difference between the two. Let's look at a couple of practical examples.

PROBLEM 6-15

You decide to paint and carpet your living room. The floor measures 15 feet by 20 feet. The ceiling is 8 feet from the floor, so the walls are all 8 feet high. The room's walls have two door openings, both 3 feet wide by 7 feet high. The room also has three wall windows, each one measuring 3 feet wide by 4 feet high. You plan to paint the walls and the ceiling, not including (obviously) the door and window openings. How many square feet of wall and ceiling area will you have to cover with paint? How many square feet of carpet will you need to completely cover the floor?

SOLUTION

Let's do the second half of this problem first. The floor measures 15 by 20 feet, so it has an area of $15 \times 20 = 300$ square feet. That's how much carpet you'll need to cover it. As for the ceiling, it has the same dimensions as the floor, so its area equals 300 square feet. Now imagine for a moment that the doors and windows didn't exist. In that case, two of the walls would have areas of $15 \times 8 = 120$ square feet, and two of the walls would have areas of $20 \times 8 = 160$ square feet. The total combined wall-and-ceiling area would then equal $(2 \times 120) + (2 \times 160) + 300 = 240 + 320 + 300 = 860$ square feet. But in reality, your room has two door openings in the walls, both of them measuring 3 by 7 feet, so they each take away $3 \times 7 = 21$ square feet from the foregoing total wall-and-ceiling area. To account for the door openings, you must subtract $21 \times 2 = 42$ from 860, leaving you with $860 - 42 = 818$ square feet. Your room also has three windows in the walls, and every one of them measures 3 by 4 feet. They each take away $3 \times 4 = 12$ square feet from the total area. To account for them, you must subtract another $12 \times 3 = 36$ square feet, leaving you with $818 - 36 = 782$ square feet. That's the amount of area that you'll have to cover with paint.

PROBLEM 6-16

You want to ship an inflated spherical ball in a cardboard box. The ball has a radius of 10 centimeters. The box's interior dimensions are 24 centimeters high, 30 centimeters wide, and 30 centimeters deep. If you place the ball in the box and then fill up all the rest of the box's interior with sawdust to serve as a packing material, how many cubic centimeters of sawdust will you need, assuming that the sawdust doesn't compress? Round the answer off to the nearest cubic centimeter.

SOLUTION

The interior volume of the box equals $24 \times 30 \times 30 = 21{,}600$ cubic centimeters. The ball has a radius of 10 centimeters, so its volume equals $4 \times 3.1416 \times 10 \times 10 \times 10 / 3 = 4189$ cubic centimeters, rounded to the nearest cubic centimeter. When you subtract the ball's volume from the box's volume, you get $21{,}600 - 4189 = 17{,}411$ cubic centimeters. That's the amount of "noncompressible sawdust" that you'll need if you want to completely fill up the box after you've placed the ball inside.

Still Struggling

You can't add an area to a volume, subtract an area from a volume, or subtract a volume from an area and ever expect to get a meaningful result. In order for sums and differences to make sense with multidimensional objects, both (or all) of the objects must have the same number of dimensions.

QUIZ

Refer to the text in this chapter if necessary. A good score is eight correct. Answers are in the back of the book.

1. **Your dining room measures exactly 5 meters square (in other words, 5 by 5 meters). You want to cover the floor with laminate tiles, each one measuring 1/4 of a meter square (1/4 meter by 1/4 meter). How many tiles will you need to cover the floor and not end up with any extra tiles?**

   A. 250
   B. 300
   C. 350
   D. 400

2. **The earth's orbit around the sun has a radius of approximately 150,000,000 kilometers. It's not a perfect circle, but it's pretty close. If we assume that the earth's orbit is indeed a perfect circle with a radius of exactly 150,000,000 kilometers, what's its circumference to the nearest million kilometers?**

   A. 471,000,000 kilometers
   B. 942,000,000 kilometers
   C. 1,480,000,000 kilometers
   D. 1,884,000,000 kilometers

3. **You and your sister own a fenced rectangular plot of land where you both want to grow vegetables for food. You want to plant potatoes, and she wants to plant tomatoes. The plot measures 500 feet wide by 1200 feet long. You and your sister agree to divide the plot in half, so you decide to put in a new fence that runs in a straight line from one corner of the rectangular plot to the opposite corner. You and your sister will then both have right-triangle-shaped plots of land. What will be the perimeter of each plot? (Here's a hint: You need the Pythagorean Theorem to work this problem out.)**

   A. 2400 feet
   B. 2700 feet
   C. 3000 feet
   D. 3500 feet

4. **What will be the interior area of each right-triangle-shaped plot of land described in Question 3?**

   A. 240,000 square feet
   B. 270,000 square feet
   C. 300,000 square feet
   D. 360,000 square feet

5. **What's the perimeter of the entire rectangular plot described in Question 3 (your half and your sister's half, combined)? In other words, how much fence did the original owner need when she enclosed the rectangular plot without dividing it?**
   A. 3400 feet
   B. 4000 feet
   C. 5000 feet
   D. 6000 feet

6. **How much total fence will the entire plot described in Question 3 contain, after you and your sister have added the new fence running diagonally from corner to corner?**
   A. 4000 feet
   B. 4700 feet
   C. 5400 feet
   D. 6700 feet

7. **The earth has a diameter (not radius!) of approximately 8000 miles. It's not a perfect sphere, but it's pretty close. If we assume that the earth is indeed a perfect sphere with a diameter of exactly 8000 miles, and if we assume that it's perfectly smooth (no mountains, hills, valleys, or other irregularities whatsoever), what's the earth's surface area to the nearest million square miles?**
   A. 201,000,000 square miles
   B. 402,000,000 square miles
   C. 603,000,000 square miles
   D. 804,000,000 square miles

8. **Once again assume that the earth constitutes a mathematically perfect, smooth sphere that's precisely 8000 miles in diameter. What's its volume to the nearest billion (thousand-million) cubic miles?**
   A. 268,000,000,000 cubic miles
   B. 603,000,000,000 cubic miles
   C. 804,000,000,000 cubic miles
   D. 1,206,000,000,000 cubic miles

9. **Imagine that you have a large cardboard box in the shape of a cube whose inside space measures 1 meter on each edge (1 meter high, 1 meter wide, and 1 meter deep). You have an unlimited supply of little sugar cubes, each of which measures 1 centimeter (1/100 of a meter) along each edge. If you stack up the sugar cubes neatly inside the box so you don't end up with any empty spaces between them, how many cubes will you need to completely fill the box with sugar?**
   A. 1,000
   B. 10,000
   C. 100,000
   D. 1,000,000

10. **Suppose that, instead of using sugar cubes to fill the box, you use tiny cubes of plastic, each of which measures 1 millimeter (1/1000 of a meter) along each edge. These cubes are too small for you to stack up by hand, but you have access to a fantastic robot that can do it for you. The robot begins stacking the tiny plastic cubes in the box, piling them up perfectly with no spaces in between. When the robot has completed the job and the box contains solid plastic, how many of those tiny plastic cubes will be in there?**
    A. 10,000,000,000
    B. 1,000,000,000
    C. 1,000,000
    D. 100,000

chapter 7

Graphs

Graphs can help us understand the relationships between changeable quantities called *variables*. When we compose a graph properly, it can reveal phenomena that we might otherwise never notice.

CHAPTER OBJECTIVES

In this chapter, you will

- Learn to show the relationship between variables on a two-axis graph.
- Compare bar graphs and pie graphs.
- Generate and interpret point-to-point graphs.
- Estimate intermediate and extended values in a graph.
- Employ curve fitting to draw the best graph for a known set of points.
- Identify trends in graphical data.
- Compare functions using a paired bar graph.

Smooth Curves

Figure 7-1 is a *continuous-curve graph* that portrays the variations in the prices of two hypothetical stocks (called X and Y) during a business day from 10:00 in the morning until 3:00 in the afternoon. Both curves represent *functions* of time, meaning that their values depend on time. On this coordinate grid, time constitutes the *independent variable* because it doesn't depend on anything else. The stock price constitutes the *dependent variable* because it depends on time. In most continuous-curve graphs, the independent variable runs along the horizontal axis, increasing from left to right, and the dependent variable runs along the vertical axis, increasing from bottom to top.

For a moment, think of the stock price as the independent variable and time as the dependent variable. To graph the situation in these terms, you can "stand the curves on their sides," as shown in Fig. 7-2. This graph seems strange, doesn't it? In what sort of financial system could time depend on the price of a stock? Obviously, when you decide to graph something, you should choose the independent variable and the dependent variable according to common sense. Figure 7-2 certainly goes against common sense! Even if you say that the vertical axis represents the independent variable and the horizontal axis represents the dependent variable, Fig. 7-2 will likely appear confusing and counterintuitive to most people.

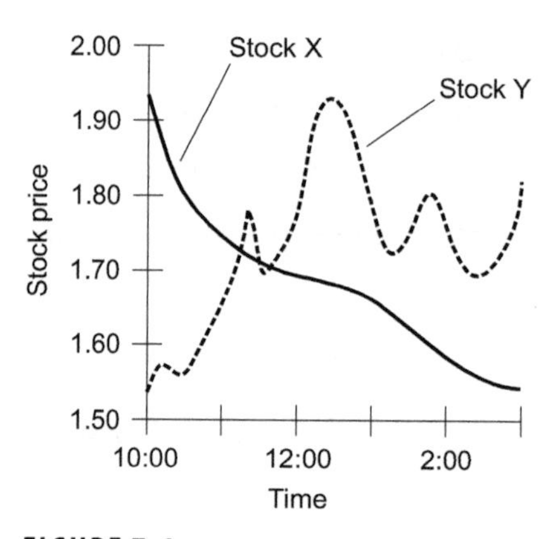

FIGURE 7-1 • The curves show fluctuations in the prices of hypothetical stocks during the course of a certain day from 10:00 in the morning until 3:00 in the afternoon.

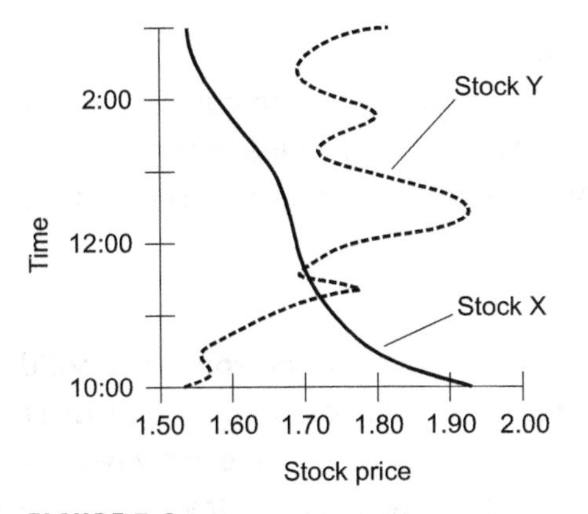

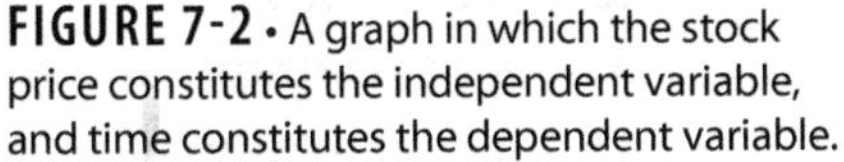
FIGURE 7-2 • A graph in which the stock price constitutes the independent variable, and time constitutes the dependent variable.

PROBLEM 7-1

Suppose that someone shows you a graph on the same grid as Fig. 7-1, and tells you that it's a plot of the price of some stock Z as a function of time. The curve shows up as a perfect circle. You sense that something's wrong with this graph. How can you explain what's wrong?

SOLUTION

Whenever you have a graph of a real-world variable, such as stock price as a function of time, the dependent variable (in this case the stock price) can never have more than one value for any given value of the independent variable (in this case time). You can't have a single stock that has more than one price at a time! If the graph appears as a circle, then at some points in time, the stock must have two different prices. That's impossible.

TIP *In a true mathematical function, the dependent variable can have* at most *one value for any particular value of the independent variable. If you see a graph where the dependent variable has more than one value for any independent-variable value, you can have complete confidence that the graph* does not *represent a true function. (Gaps, where you see no dependent-variable value at all, are okay.)*

PROBLEM 7-2

Imagine that, in Fig. 7-1, you "stand the curves on their sides" so that they mimic the curves in Fig. 7-2, but you leave time on the horizontal axis and stock price on the vertical axis. Will the resulting curves represent a realistic scenario?

SOLUTION

One of them will, and one of them won't. The solid curve, representing stock X, will represent a realistic situation because stock X will never have more than one price at any given time. However, the dashed curve, representing stock Y, will make no sense. It will have more than one price at certain times. In fact, if you draw this graph and then analyze it closely, you'll find that stock Y will have as many as *seven* different prices at a time.

Bars and a Pie

In a *vertical bar graph*, the independent variable goes along a horizontal axis increasing from left to right, and the dependent variable goes along a vertical axis increasing from bottom to top. Specific function values appear as the heights of rectangles called *bars*, all equally wide. (The bar widths don't have mathematical significance. We choose the left-to-right bar widths for clarity and ease of reading, making all of the bars equally wide.) Figure 7-3 is a vertical bar graph of the price of stock Y at intervals of one hour. Figure 7-3 reveals less detail than Fig. 7-1 does, but some people find bar graphs easier to read than continuous-curve graphs.

In a *horizontal bar graph*, we show the independent variable along the vertical axis increasing from bottom to top, and the dependent variable along the horizontal axis increasing from left to right. We plot function values as the widths of bars having equal heights (which we choose only for clarity of illustration; the bar heights have no mathematical meaning). Figure 7-4 is a horizontal bar graph of the price of stock Y at intervals of one hour. Horizontal bar graphs are rarely used because readers tend to get the independent and dependent variables confused.

A *pie graph* is a circular graph that does a good job of showing relative proportions or percentages. For example, consider the "letter grades" (A, B, C, D, or F) that the students in a class receive when they take a hypothetical test.

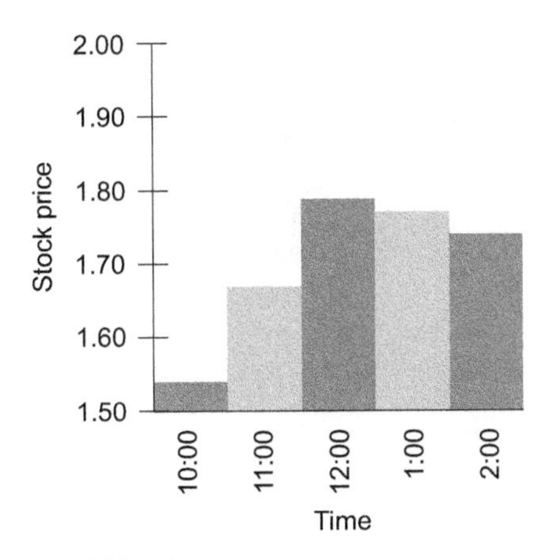

FIGURE 7-3 • A vertical bar graph of stock price versus time. The independent variable runs along the horizontal axis. The dependent variable runs along the vertical axis.

Figure 7-5A shows the results as a vertical bar graph. Figure 7-5B illustrates the same situation as a pie graph. The size of any particular "pie slice" is directly proportional to the percentage of students in the class receiving the corresponding grade. In a pie graph, the dependent-variable samples always add up to 100%, just as, in a real pie, the sum of all the slices adds up to the whole pie—no more and no less.

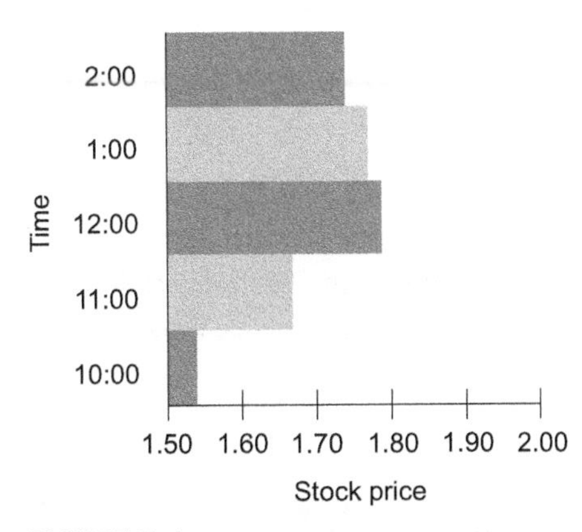

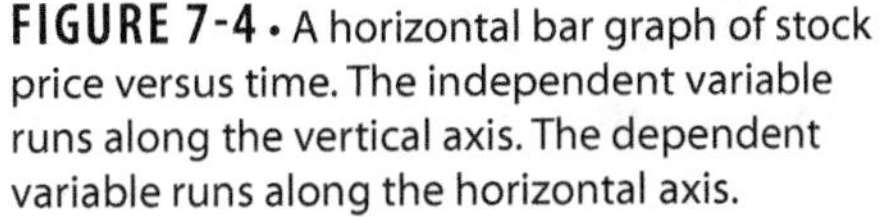
FIGURE 7-4 • A horizontal bar graph of stock price versus time. The independent variable runs along the vertical axis. The dependent variable runs along the horizontal axis.

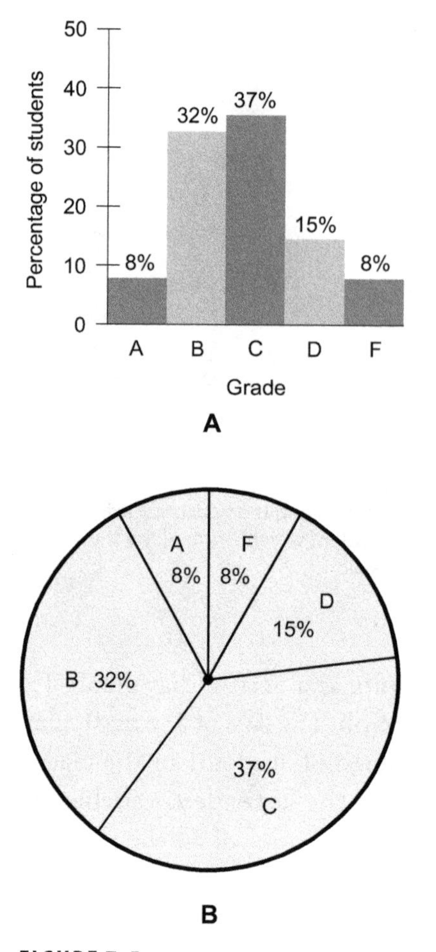

FIGURE 7-5 • At A, a bar graph showing student grades. At B, a pie graph showing the same situation.

PROBLEM 7-3

Figure 7-6 shows the percentage of the work force in a certain city that "calls in sick" on each day during a particular work week. What, if anything, is wrong with this graph?

SOLUTION

First of all, this figure shows a horizontal bar graph. We'd do better to portray the data as a vertical bar graph, with the day of the week running along the horizontal axis and the percentage of the work force following the vertical axis. Secondly, even if we accept Fig. 7-6 as it is, the horizontal

(dependent-variable) scale covers an unnecessarily large range, so that the values in the graph are hard to read. We'd find the graph easier to read, as well as more accurate, if the horizontal scale showed values only in the range of, say, 0% to 10%. We could further improve this graph by writing down the actual percentage numbers at the end of each bar.

TIP *Whenever we compose a graph, we should choose sensible scales for the dependent and independent variables. If either scale spans a range of values much greater than necessary, the* resolution *(detail) of the graph will suffer. If either scale doesn't have a large enough span, then we won't have enough room to show the entire function; some legitimate and important values or points will get "cut off."*

PROBLEM 7-4

What's going on with the percentage values depicted in Fig. 7-6? The values don't add up to 100%. Shouldn't they?

SOLUTION

Not necessarily in this situation. If all the values did add up to 100%, we'd have a coincidence—and a bad omen, too, perhaps indicating a disease epidemic! In fact, during a serious epidemic, we might actually see the values add up to more than 100% for the week. If everybody showed up for work every day for the whole week, the sum of the percentages would equal 0%, and the graph would appear blank, containing no bars at all.

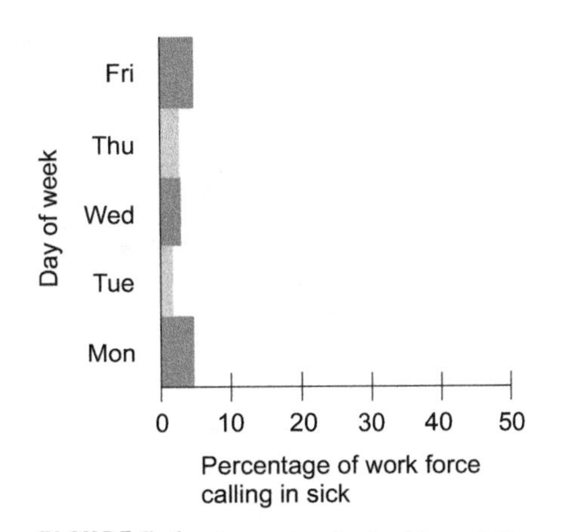

FIGURE 7-6 • Illustration for Problems 7-3 and 7-4 and their solutions.

Point-to-Point Graphs

When we draw a *point-to-point graph*, we use axes like the ones in continuous-curve graphs, such as Fig. 7-1, but we put down actual values only for a few selected points, connecting those points with a jagged line rather than a smooth curve. We get less precision than we do with the continuous-curve graph, but we don't have to do as much work to gather the data.

In the point-to-point graph of Fig. 7-7, we plot the price of stock Y (from the situation we've worked with so far in this chapter) every half hour from 10:00 in the morning to 3:00 in the afternoon. The resulting jagged line doesn't exactly show the stock prices at the in-between times. Nevertheless, it gives us a fair idea of how the stock price varies as the day goes by.

TIP ***Whenever we plot a point-to-point graph, we must include a certain minimum number of points, and we must make sure that they lie close enough together. If a point-to-point graph showed the price of stock Y at hourly intervals, it wouldn't come as close as Fig. 7-7 does to representing the actual stock-price function. If a point-to-point graph showed the price at 15-minute intervals, it would come closer than Fig. 7-7 does to the moment-to-moment stock-price function.***

PROBLEM 7-5

Draw a point-to-point graph of the price of stock Y as a function of time, including only the actual values on the hour from 10:00 a.m. to 3:00 p.m.

SOLUTION

Figure 7-8 shows the result. As you can see, it's less precise than Fig. 7-7.

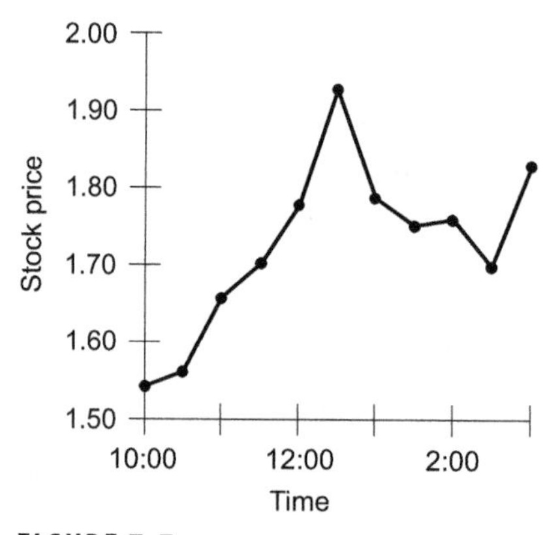

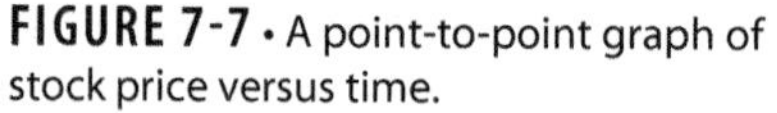
FIGURE 7-7 • A point-to-point graph of stock price versus time.

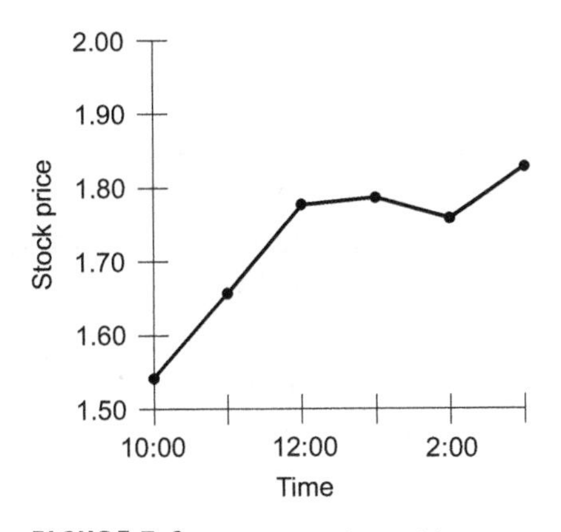

FIGURE 7-8 • Illustration for Problem 7-5 and its solution.

PROBLEM 7-6

Draw a *dual point-to-point graph* that illustrates the results of Figs. 7-7 and 7-8, superimposed on each other to emphasize the difference between the two.

SOLUTION

Figure 7-9 shows the result. The dashed jagged line shows the graph based on half-hourly intervals. The solid jagged line shows the graph based on hourly intervals.

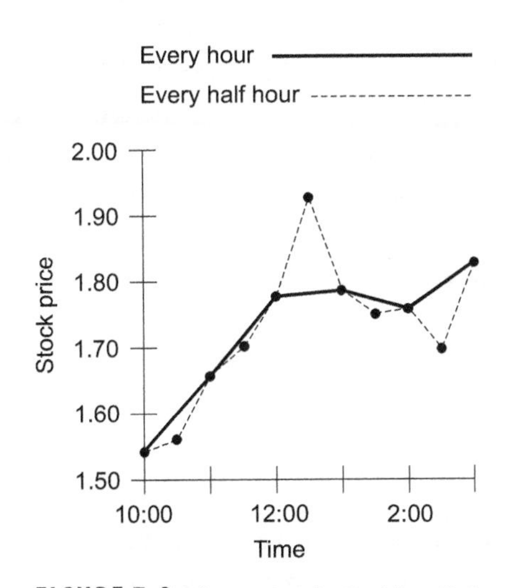

FIGURE 7-9 • Illustration for Problem 7-6 and its solution.

Interpolation

The term *interpolate* means "put between" or "estimate the values between." When confronted with an incomplete graph, we can sometimes insert estimated data in the gap(s) to make the graph appear complete. Figure 7-10 shows a graph of the price of our hypothetical stock Y, but a gap occurs during the noon hour. We don't know exactly what happened to the stock price during that hour, but we can fill in the graph using *linear interpolation*. We simply draw a straight line between the end points of the gap.

Linear interpolation usually produces an inaccurate result. But sometimes we're better off with an approximation, even a crude one, than we are with no data at all. When we compare Fig. 7-10 with Fig. 7-1, we see a considerable *linear interpolation error* for this particular gap in the graph for stock Y. Our estimate overlooks the large "hump" that the curve follows in Fig. 7-1, indicating a surge in the stock price of stock Y during the noon hour. Interpolation offers convenience—but a considerable risk of error.

Still Struggling

You might ask, "Can a function exist for which linear interpolation fills in a gap without any error?" The answer is yes. If you know that the graph of a function is a straight line, then you can use linear interpolation to fill in a gap, and the result will have no error. Consider, for example, a car that accelerates at a constant rate. If its speed-versus-time graph appears as a perfectly straight line with a small gap, then you can employ linear interpolation to determine the car's speed at points inside the gap as shown in Fig. 7-11. In this graph, the black dashed line represents the actual measured data, and the gray solid line represents the interpolated data.

PROBLEM 7-7

Consider once again the graph of stock Y as a function of time, expressed as a continuous curve. Suppose that a gap exists in the data between 10:30 a.m. and

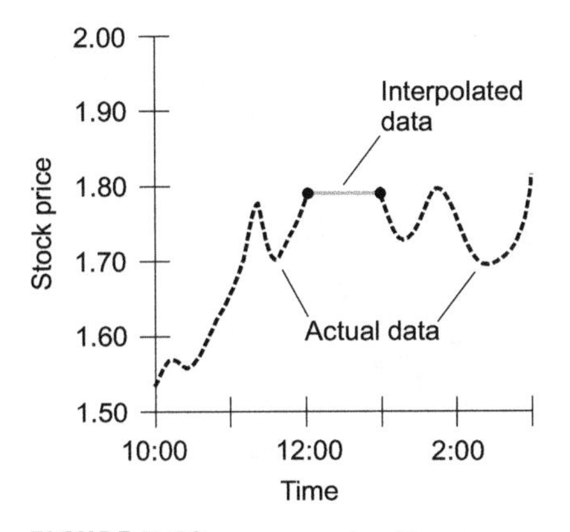

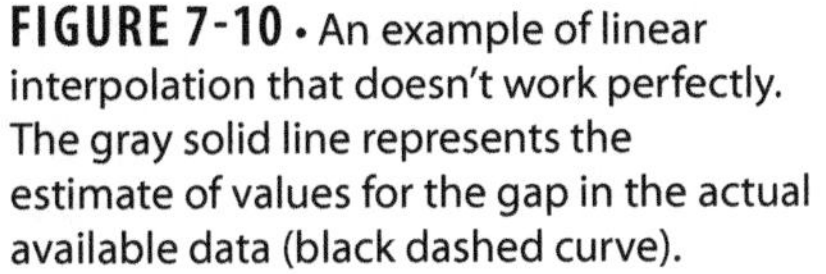
FIGURE 7-10 • An example of linear interpolation that doesn't work perfectly. The gray solid line represents the estimate of values for the gap in the actual available data (black dashed curve).

11:00 a.m. Use linear interpolation to fill in this gap. How closely does the result approximate the actual curve as it appears in Fig. 7-1?

SOLUTION

Figure 7-12 shows the curve with the gap, along with the straight line for the linear interpolation. This estimate comes quite close to the actual path of the curve, as we can see when we compare it with Fig. 7-1. The error in this case is far smaller than the error in the estimate of values between 12:00 noon and 1:00 p.m. in Fig. 7-10.

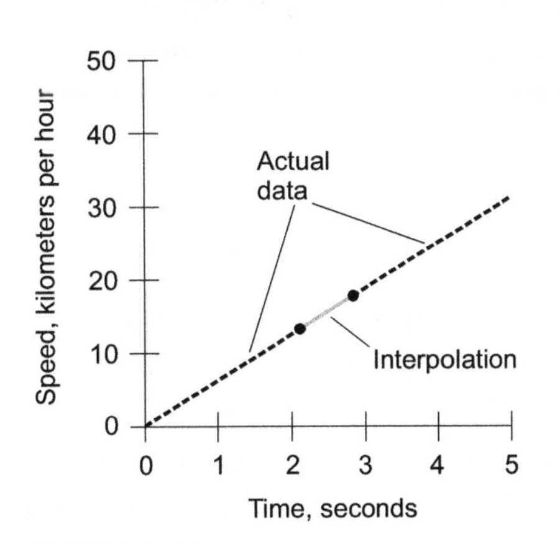

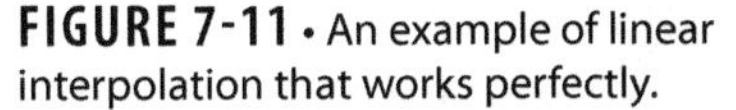
FIGURE 7-11 • An example of linear interpolation that works perfectly.

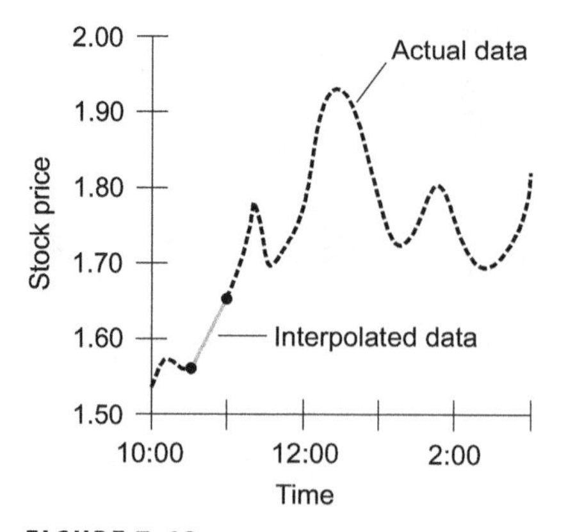

FIGURE 7-12 • Illustration for Problem 7-7 and its solution.

PROBLEM 7-8

How does the extent of the error in linear interpolation vary with the size of the data gap for any given curve?

SOLUTION

In general, the linear interpolation error decreases as the gap gets narrower. For example, in the graph of either stock X or stock Y, as shown in Fig. 7-1, we should expect that linear interpolation would give us a better result, in general, for a gap of 30 minutes than for a gap of an hour. We should expect even better results, in general, for a gap of 20 minutes, and still better results for a gap of only 10 minutes.

Curve Fitting

Curve fitting allows us to approximate a point-to-point graph, or fill in a graph containing one or more gaps, so that we can make it look like a continuous curve. Figure 7-13 is an approximate graph of the price of our imaginary stock Y, based on points determined at intervals of 30 minutes as generated by curve fitting. Here, the dashed black line shows the actual moment-to-moment stock price, and the solid gray line shows a fitted curve based on prices sampled at 30-minute intervals. The fitted curve doesn't

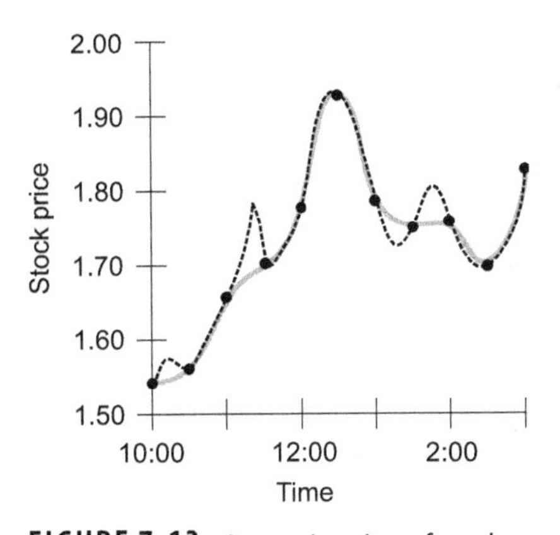

FIGURE 7-13 • Approximation of stock price as a continuous function of time with curve fitting. The solid gray curve represents the approximation. The black dashed curve represents the actual stock price.

precisely represent the actual stock price at every instant, but it comes close most of the time.

As we determine and plot the values at more and more frequent intervals (say every 20 minutes, then every 15 minutes, then every 10 minutes), we find that curve fitting gives us better and better results. When we determine and plot the values infrequently, however, curve fitting can introduce large errors, just as linear interpolation can do. Figure 7-14 shows an example in which we sample the actual price of stock Y at hourly intervals. Curve fitting does a poor job in this case.

PROBLEM 7-9

Use curve fitting to approximate the graph of stock Y, based on points determined at intervals of 15 minutes. How does this result compare with the 30-minute-interval graph of Fig. 7-13?

SOLUTION

Figure 7-15 illustrates this graph. The approximation (gray curve) closely follows the actual data (dashed curve), except in the interval between 11:15 a.m. and 11:30 a.m.

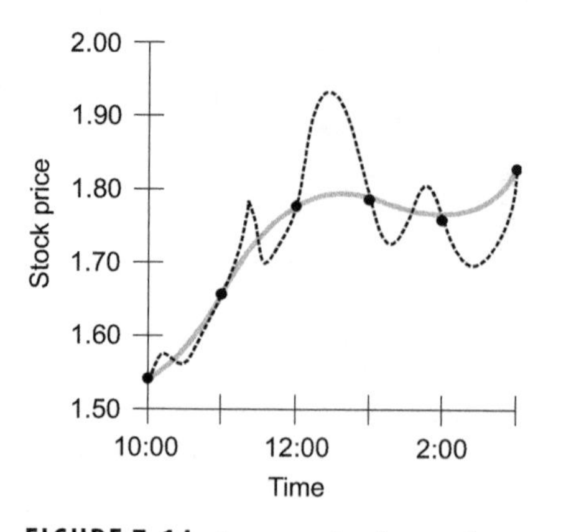

FIGURE 7-14 • An example of curve fitting in which insufficient data samples exist. The solid gray line represents the approximation. The black dashed curve represents the actual stock price.

PROBLEM 7-10

How does the extent of the error in curve fitting vary with the size of the intervals between the data points?

SOLUTION

In general, the curve-fitting error decreases as the data points get closer together. When we compare Figs, 7-13, 7-14, and 7-15, we can see this effect.

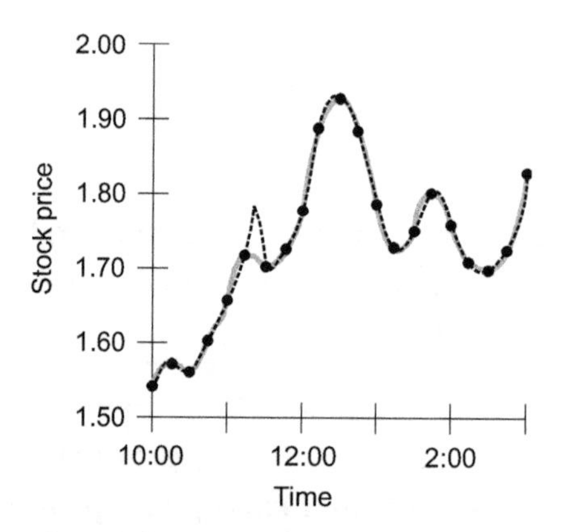

FIGURE 7-15 • Illustration for Problem 7-9 and its solution.

Extrapolation

The term *extrapolate* means "put outside of" or "estimate the values outside of." When a function has a continuous-curve graph with time as the independent variable, *extrapolation* means the same thing as *short-term forecasting*. Figure 7-16 shows two examples.

In Fig. 7-16A, we plot the price of our make-believe stock X as a function of time until 2:00 p.m., and then we try to forecast its price for 1 hour into the future, based on its past performance. In this case, we use *linear extrapolation*, the simplest extrapolation technique. We project the curve ahead as a straight

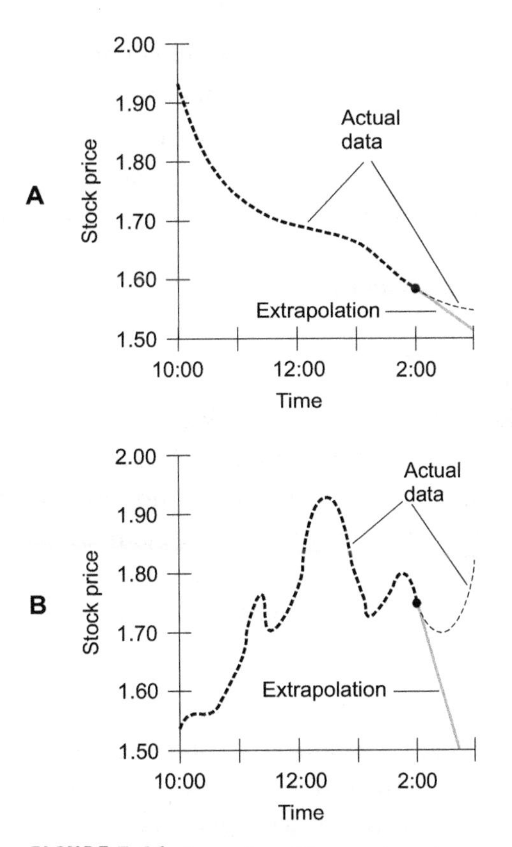

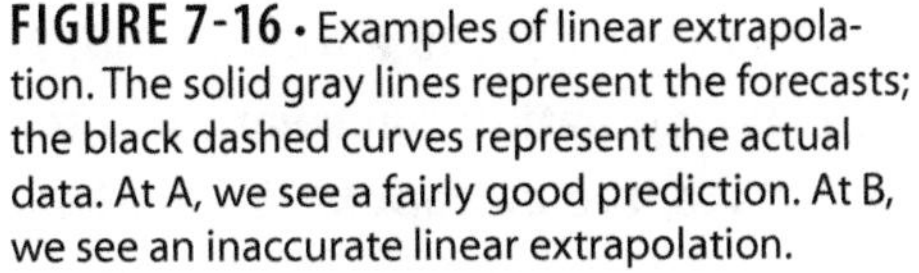

FIGURE 7-16 • Examples of linear extrapolation. The solid gray lines represent the forecasts; the black dashed curves represent the actual data. At A, we see a fairly good prediction. At B, we see an inaccurate linear extrapolation.

line. Compare this graph with the solid curve in Fig. 7-1. In this case, linear extrapolation works fairly well.

Figure 7-16B shows the price of stock Y plotted until 2:00 p.m. Then we employ linear extrapolation in an attempt to predict its behavior for the next hour. As we can see by comparing this graph with the dashed curve in Fig. 7-1, linear extrapolation doesn't do a good job at all in this situation!

Some graphs lend themselves to extrapolation, while other graphs don't. In general, as a curve becomes more complicated, extrapolation grows less precise, and more subject to large errors, such as we see in Fig. 7-16B. Also, as the extent (distance) of extrapolation increases for a given curve, the accuracy decreases, so we can expect the results to differ more and more from actual outcomes as we move away from the last known data point.

TIP ***If you want to take advantage of extrapolation, you can use computer software written for that purpose. Machines can "see" subtle characteristics of functions that humans overlook. Hurricane forecasters make use of extrapolation programs in an attempt to predict where, or if, a storm will make landfall. The most sophisticated programs can predict future curve trends (as opposed to doing simple linear extrapolation). But no program is completely infallible!***

PROBLEM 7-11

Suppose that we know the data for the price of stock Y at 15-minute intervals from 11:00 a.m. to 3:00 p.m. We plot those points and create an estimate of the price-versus-time graph by means of curve fitting. Then we attempt to extrapolate backward in time from 11:00 a.m. to 10:00 a.m. by extending the graph as a straight line. What does our graph look like? How does it compare with the actual price of stock Y from 10:00 a.m. to 3:00 p.m. as shown in Fig. 7-1?

SOLUTION

Figure 7-17 shows the result. The black dots represent the data points for the price of stock Y at 15-minute intervals from 11:00 a.m. to 3:00 p.m. The solid gray curve shows the estimate of the price-versus-time graph for stock Y by means of curve fitting. (The fitted curve contains the same significant error between 11:15 a.m. and 11:30 a.m. that Fig. 7-15 has.) The dashed gray line shows the extrapolation backward in time from 11:00 a.m. to 10:00 a.m. As we can see, the extrapolation works okay from 11:00 a.m. back to 10:30 a.m., but not so well from 10:30 a.m. back to 10:00 a.m.

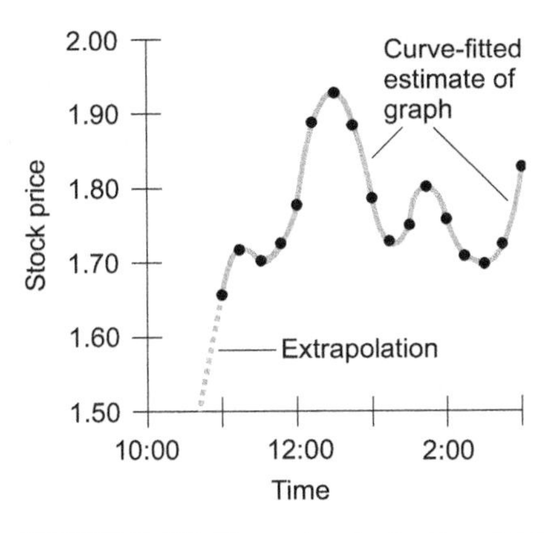

FIGURE 7-17 • Illustration for Problem 7-11 and its solution.

PROBLEM 7-12

Imagine that we know the data for the price of stock Y at 15-minute intervals from 10:00 a.m. to 2:45 p.m. We plot the points and estimate the graph with curve fitting. Then we try to extrapolate forward in time from 2:45 p.m. to 3:00 p.m. with a straight line. What does our graph look like? How does it compare with the actual price of stock Y from 10:00 a.m. to 3:00 p.m. as shown in Fig. 7-1?

SOLUTION

See Fig. 7-18. The black dots show the data points for the price of stock Y at 15-minute intervals from 10:00 a.m. to 2:45 p.m. The solid gray curve shows the estimate of the price-versus-time graph for stock Y by means of curve fitting. (The fitted curve contains the same significant error between

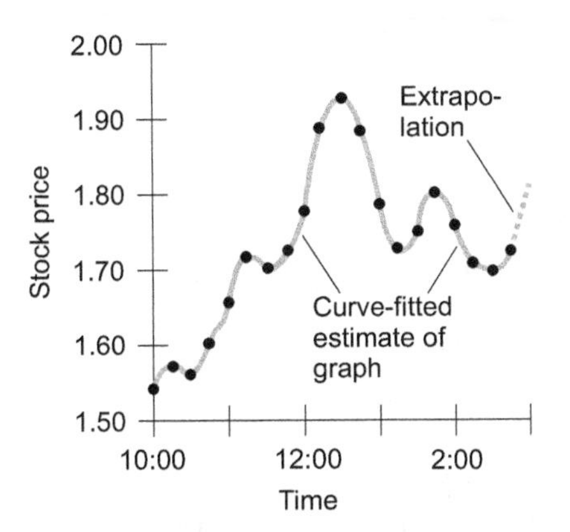

FIGURE 7-18 • Illustration for Problem 7-12 and its solution.

11:15 a.m. and 11:30 a.m. that Fig. 7-15 has.) The dashed gray line shows the extrapolation forward in time from 2:45 p.m. to 3:00 p.m. As we can see, the extrapolation works okay all the way through this 15-minute interval.

Trends

Functions of time (or anything else) can fluctuate in an infinite variety of ways. However, some functions exhibit specific behaviors called *trends*. In the real world, you'll encounter trends of four different types. Let's define each "flavor" of trend, looking at four hypothetical stocks called Q, R, S, and T as they vary in price between 10:00 a.m. and 3:00 p.m. on a certain day.

1. We call a function *strictly nonincreasing* when the dependent variable *never* grows larger (or more positive) as the independent variable grows larger (or more positive). The dashed curve in Fig. 7-19 shows the behavior of a hypothetical stock Q, whose price never rises between 10:00 a.m. and 3:00 p.m. This function is nonincreasing over the interval.
2. We call a function *strictly nondecreasing* when the dependent variable *never* grows smaller as the independent variable grows larger. The solid curve in Fig. 7-19 portrays the behavior of a hypothetical stock R, whose price never falls between 10:00 a.m. and 3:00 p.m. This function is nondecreasing over the interval.
3. We say that a function trends *generally downward* when the dependent variable grows smaller *on the average*, but not necessarily at every point,

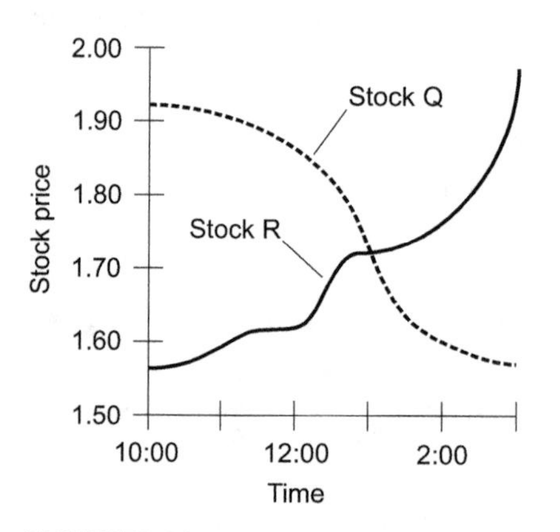

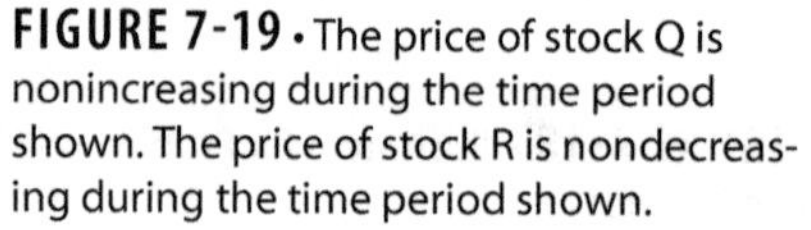
FIGURE 7-19 • The price of stock Q is nonincreasing during the time period shown. The price of stock R is nondecreasing during the time period shown.

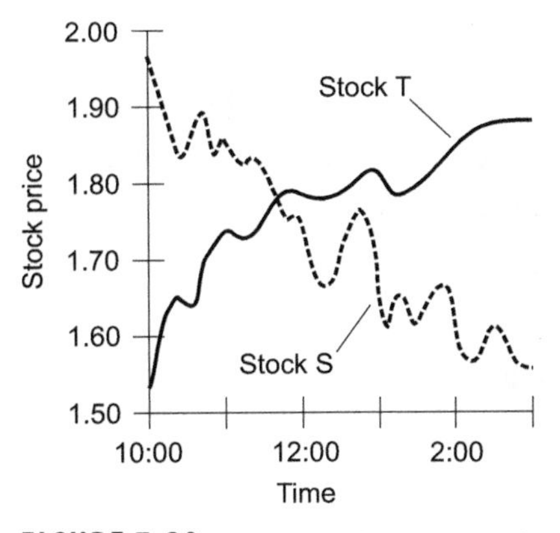

FIGURE 7-20 • The price of stock S trends generally downward during the time period shown. The price of stock T trends generally upward during the time period shown.

as the independent variable grows larger. The dashed curve in Fig. 7-20 portrays the behavior of a hypothetical stock S, whose price trends generally downward between 10:00 a.m. and 3:00 p.m.

4. We say that a function *trends generally upward* when the dependent variable grows larger *on the average*, but not necessarily at every point, as the independent variable grows larger. The solid curve in Fig. 7-20 portrays the behavior of a hypothetical stock T, whose price trends generally upward between 10:00 a.m. and 3:00 p.m.

Still Struggling

Whenever you identify a function as having a certain "trend flavor" (one of the four types described above), you should specify the interval over which you claim that the characteristic applies. In Figs. 7-19 and 7-20, you can say certain things about the four stocks Q, R, S, and T only between 10:00 a.m. and 3:00 p.m. on a certain day. These graphs don't tell you what happens (or happened) to the prices of any of those stocks after 3:00 p.m. on that day, or before 10:00 a.m. on that day, although you might feel tempted to extrapolate in an attempt to "predict" events after 3:00 p.m. and then assign an "expected future trend" on that basis. That's what real-time stock traders do. Sometimes they make money that way, and sometimes they don't.

PROBLEM 7-13

Figure 7-21 shows a hypothetical stock called U as a function of time during a certain day. Over what interval (or intervals) *exceeding one continuous hour* can we say that the price of stock U is

- **strictly nonincreasing?**
- **strictly nondecreasing?**
- **trending generally downward?**
- **trending generally upward?**

SOLUTION

It appears that no continuous hourly intervals exist during which the price of stock U is strictly nonincreasing, except *maybe* for the period starting a few minutes before 10:00 a.m. and ending a few minutes before 11:00 a.m. The price is strictly nondecreasing over the entire time interval from slightly before 12:00 noon all the way until 5:00 p.m. The price trends generally downward from 9:00 a.m. until a few minutes before 12:00 noon, and trends generally upward from then until 5:00 p.m.

Paired Bar Graphs

Imagine two cities, Happyton and Blissville, located far from each other in North America. We plan to move from Happyton to Blissville. People tell us

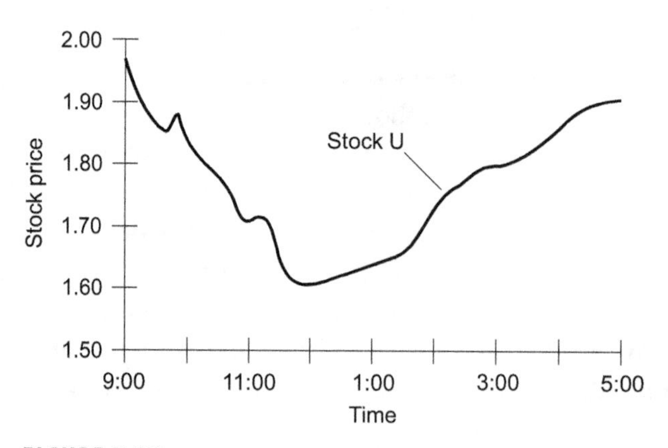

FIGURE 7-21 • Illustration for Problem 7-13 and its solution.

that Happyton has wet summers and dry winters, while in Blissville we should expect dry summers and wet winters. People also warn us that the average temperature varies drastically between summer and winter in Happyton (as we know, having dwelt there for years), but varies much less between summer and winter in Blissville.

We go to the Internet and gather data about the two towns. We find a collection of tables showing the average monthly temperature in degrees Celsius and the average monthly rainfall in centimeters, gathered over the past several decades for many places throughout the world, including Happyton and Blissville. Table 7-1 shows the average monthly temperature and rainfall for Happyton over the past 20 years. Table 7-2 shows the average monthly temperature and rainfall for Blissville over the past 20 years.

Let's graphically compare the average monthly temperature and the average monthly rainfall for Happyton. Figure 7-22 is a *paired bar graph* showing the average monthly temperature and rainfall there. We base this graph on the data from Table 7-1. The horizontal axis has 12 intervals, each one showing a month of the year. Time constitutes the independent variable. The left-hand vertical

TABLE 7-1 Average monthly temperature and rainfall data for the imaginary town of Happyton.

Month	Average Temperature in Degrees Celsius	Average Rainfall in Centimeters
January	2.1	0.4
February	3.0	0.2
March	9.2	0.3
April	15.2	2.8
May	20.4	4.2
June	24.9	6.3
July	28.9	7.3
August	27.7	8.5
September	25.0	7.7
October	18.8	3.6
November	10.6	1.7
December	5.3	0.5

TABLE 7-2 Average monthly temperature and rainfall data for the imaginary town of Blissville.

Month	Average Temperature in Degrees Celsius	Average Rainfall in Centimeters
January	10.5	8.1
February	12.2	8.9
March	14.4	6.8
April	15.7	4.2
May	20.5	1.6
June	22.5	0.4
July	23.6	0.2
August	23.7	0.3
September	20.7	0.7
October	19.6	2.4
November	16.7	3.4
December	12.5	5.6

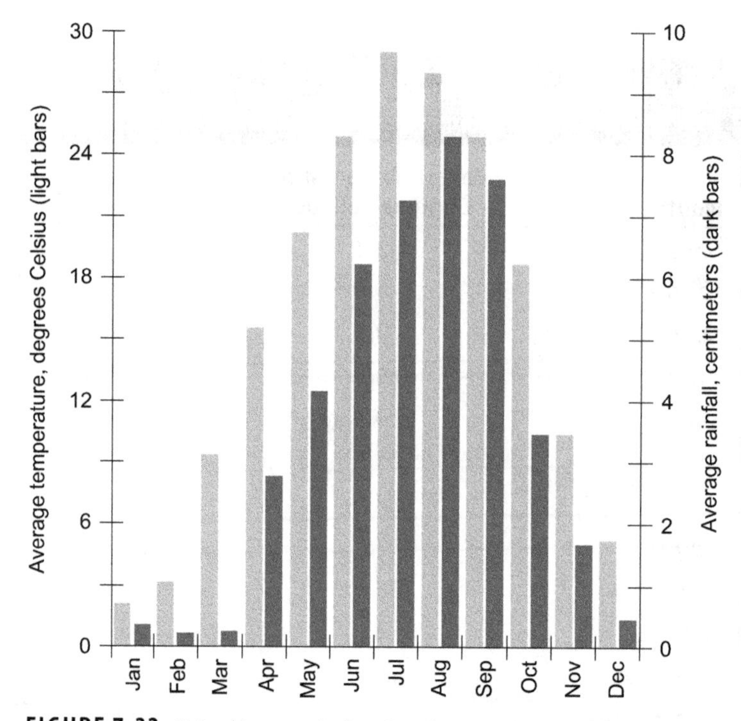

FIGURE 7-22 • Paired bar graph showing the average monthly temperature and rainfall for the imaginary town of Happyton.

scale portrays the average monthly temperatures, and the right-hand vertical scale portrays the average monthly rainfall amounts. Both of these are dependent variables (functions of the time of year). We portray the average monthly temperatures as light gray bars, and the average monthly rainfall amounts as dark gray bars. We can see that the temperature and rainfall both follow a pattern. In general, the warmer months have more precipitation than the cooler months in Happyton.

Now let's make a similar comparison for Blissville. Figure 7-23 is a paired bar graph showing the average monthly temperature and rainfall there, based on the data from Table 7-2. We can see that the temperature varies less between winter and summer in Blissville than it does in Happyton. But the rainfall profile for Blissville, as a function of the time of year, differs drastically from the rainfall profile for Happyton. The winters in Blissville, especially the months of January and February, have a lot of precipitation, while the summers, particularly June, July, and August, get almost none. We can infer this information if we scrutinize the tabular data at length, but we can see it immediately when we glance at the graphs.

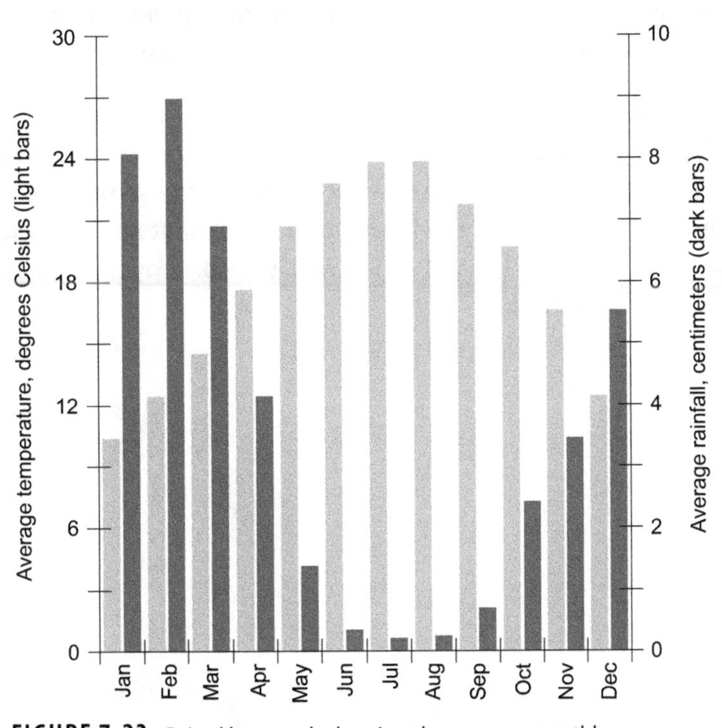

FIGURE 7-23 • Paired bar graph showing the average monthly temperature and rainfall for the imaginary town of Blissville.

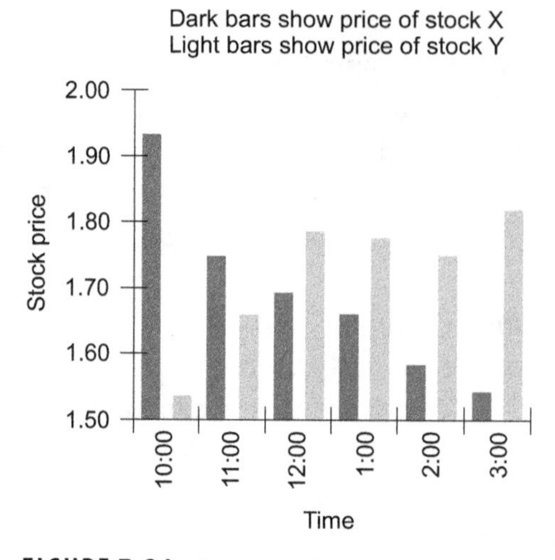

FIGURE 7-24 • Illustration for Problem 7-14 and its solution.

PROBLEM 7-14

Plot the prices of the hypothetical stocks X and Y, which we evaluated earlier in this chapter (and graphed in Fig. 7-1), at specific time points on the "top of each hour" from 10:00 to 3:00, in the form of a paired bar graph.

SOLUTION

See Fig. 7-24. The dark bars show the price of stock X at the top of each hour. The light bars show the price of stock Y at the same times. Note that in this graph, the "tops of the hours" don't appear at marks along the horizontal axis as they do in Fig. 7-1. Instead, they occupy the spaces between pairs of adjacent marks.

QUIZ

Refer to the text in this chapter if necessary. A good score is eight correct. Answers are in the back of the book.

1. **Suppose that someone shows you a continuous-curve graph portraying the temperature during a certain day (from midnight to the next midnight) "as a function of time." You look at the graph and know that it can't represent a function because**
   A. certain temperatures appear to occur more than once.
   B. two different temperatures appear to occur at certain times.
   C. it trends generally downward for the entire period of time.
   D. it's strictly nondecreasing for the entire period of time.

2. **Which of the following characteristics does a continuous-curve graph *never* have?**
   A. The independent variable runs along the horizontal axis, increasing in value as we move toward the right.
   B. The dependent variable runs along the vertical axis, increasing in value as we move upward.
   C. It offers more precision than a point-to-point graph showing the same function.
   D. It reveals less detail than a vertical bar graph showing the same function.

3. **Suppose that we have collected some data for the temperature versus time over the course of a certain day from 6:00 in the morning until 6:00 in the evening. Which of the following graphs will give us the most precise rendition of the actual function?**
   A. A graph of data taken at 15-minute intervals and filled in by curve fitting
   B. A vertical bar graph made from data taken at hourly intervals
   C. A point-to-point graph made from data taken at 30-minute intervals
   D. A pie graph made from data taken at hourly intervals

4. **Look again at the paired-bar graphs of Figs. 7-22 and 7-23. Which of the observed phenomena trend generally upward over the first eight months of the year (from January through August)?**
   A. The average monthly rainfall for Happyton
   B. The average monthly temperature for Happyton
   C. The average monthly temperature for Blissville
   D. All of the above

5. **In a horizontal bar graph, the left-to-right widths of the bars depend on the value of**
   A. the independent variable.
   B. the dependent variable.
   C. the function at a specific point.
   D. nothing in particular.

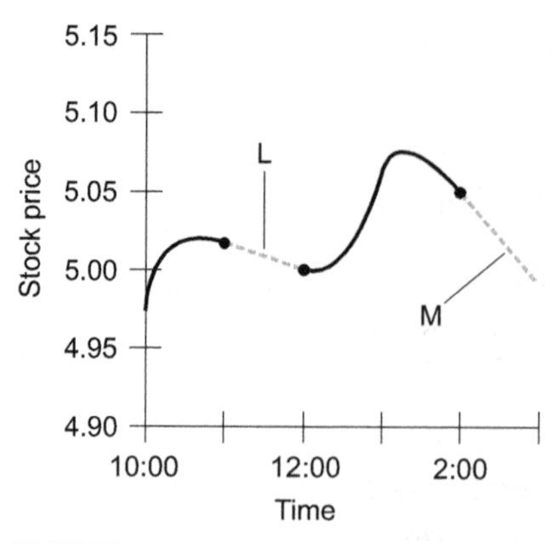

FIGURE 7-25 • Illustration for Quiz Questions 9 and 10.

6. **In a vertical bar graph, the left-to-right widths of the bars depend on the value of**
   A. the independent variable.
   B. the dependent variable.
   C. the function at a specific point.
   D. nothing in particular.

7. **Which of the following characteristics does a point-to-point graph *never* have?**
   A. The independent variable runs along the horizontal axis, increasing in value as we move toward the right.
   B. The dependent variable runs along the vertical axis, increasing in value as we move upward.
   C. It offers more precision than a continuous-curve graph showing the same function.
   D. It reveals more detail than a horizontal bar graph showing the same function.

8. **The sample proportions *always* add up to 100% in a**
   A. pie graph.
   B. continuous-curve graph.
   C. horizontal bar graph.
   D. paired bar graph.

9. **Figure 7-25 shows the price of a hypothetical stock over the course of a certain day. The dashed line marked L illustrates an example of**
   A. curve fitting.
   B. extrapolation.
   C. interpolation.
   D. trending.

10. **In Fig. 7-25, the dashed line marked M illustrates an example of**
   A. curve fitting.
   B. extrapolation.
   C. interpolation.
   D. trending.

chapter 8

Finances

If managing money were easy, everyone would be rich. Most of us have a long way to go before we reach that ideal. In this chapter, you'll review the basic mathematics of some important aspects of personal finance.

CHAPTER OBJECTIVES

In this chapter, you will

- Learn how to count and make change without having to do any subtraction.
- Find out how interest on savings can profit you over time.
- See how interest on loans can cost you over time.
- Discover how loan amortization works.
- Learn how common taxation schemes work, and how they can affect you.

Dollars and Change

Whenever you express cash amounts in U.S. dollars down to the penny, you must put the decimal point, which separates dollars from cents, two digits from the far right. In other words, the decimal point must always have two digits after it. If you need to do calculations with large amounts of money, you can insert commas at multiples of three digits to the left of the decimal point. You should always precede an American cash figure by either a dollar sign ($) or by the capital letters USD (an abbreviation of "United States Dollars").

When you want to express cash amounts less than one dollar, you can place the decimal point immediately after the dollar sign or the letters USD, and then write down the two digits representing the number of cents. Alternatively, you can write down the dollar sign or USD, then a cipher (numeral zero), then the decimal point, and then the number of cents. For example, someone might say that your new house, with all new appliances and a complete paint job, after all realtor's commissions and taxes and fees and other various middle-person's "takes" have been deducted, will cost you USD 273,868.43 (two hundred seventy-three thousand, eight hundred sixty-eight dollars and forty-three cents). A single "touch-up" paint brush, in contrast, might set you back by only $.99 or $0.99 (ninety-nine cents).

Once in awhile, people leave out the pennies when denoting large cash amounts. You might rather say that your new house cost USD 273,868, rounding off to the nearest dollar. When the values grow into the millions, billions (thousand-millions) or trillions (million-millions) of dollars, some people round off to the nearest million, billion, or trillion, and write the word itself instead of the zeros afterward. So, for example, a senior corporate executive might eagerly await her year-end bonus of $15 million; your state government might run a deficit of USD 5 billion this year; a nation might have a total debt of USD 25 trillion by the end of this year, accumulated over the course of its history.

TIP ***When you write down any amount of cash less than a dollar, always make the decimal point easy to see, so your reader doesn't think you mean, for example, ninety-nine dollars ($99) instead of ninety-nine cents ($.99 or $0.99)!***

Still Struggling

When you add and subtract cash amounts, always remember to include all the zeros, whether they occur before the decimal point or after it. If you forget any of those ciphers, you could end up thinking that you have more or less cash than you actually have—maybe a lot more or less.

When a competent sales person makes change, he'll use addition to check the subtraction when you pay with paper currency and expect cash back. Suppose that you buy something for $3.47 and use a $100.00 bill for payment. Subtraction will tell you that you should get $96.53 in change. However, the salesperson can work out your refund without having to subtract at all. He'll count your change back and recite the figures. In this case, he might go through the motions and announcements in order, as follows:

- Hand you three pennies and say "three dollars and forty-eight, forty-nine, fifty"
- Hand you two quarters and say "seventy-five, four dollars"
- Hand you a $1.00 bill and say "five"
- Hand you a $5.00 bill and say "ten"
- Hand you two $20.00 bills and say "thirty, fifty"
- Hand you a $50.00 bill and say "one hundred dollars"

At this point you can have a little fun and say, "I'm sorry to trouble you, but can you give me a twenty ($20.00 bill), two tens ($10.00 bills) and two fives ($5.00 bills) for that fifty ($50.00 bill)? The salesperson will look at you for a moment, take back the $50.00 bill, and then go through the following routine:

- Hand you a $20.00 bill and say "twenty"
- Hand you two $10.00 bills and say "thirty, forty"
- Hand you two $5.00 bills and say "forty-five, fifty dollars"

PROBLEM 8-1

Describe an alternative way to make change in the foregoing situation.

SOLUTION

The salesperson might give you back the change for a $100.00 bill after you make a $3.47 purchase by going through the following routine:

- Hand you three pennies and say "three dollars and forty-eight, forty-nine, fifty"
- Hand you two quarters and say "seventy-five, four dollars"
- Hand you six $1.00 bills and say "five, six, seven, eight, nine, ten"
- Hand you three $10.00 bills and say "twenty, thirty, forty"
- Hand you three $20.00 bills and say "sixty, eighty, one hundred dollars"

PROBLEM 8-2

Describe a way that the salesperson can make change in the foregoing situation if he has no paper notes larger than a $10.00 bill.

SOLUTION

If the salesperson doesn't have any bills larger than a "10-spot," he might count out your change as follows:

- Hand you three pennies and say "three dollars and forty-eight, forty-nine, fifty"
- Hand you two quarters and say "seventy-five, four dollars"
- Hand you a $1.00 bill and say "five"
- Hand you a $5.00 bill and say "ten"
- Hand you nine $10.00 bills and say "twenty, thirty, forty, fifty, sixty, seventy, eighty, ninety, one hundred dollars"

Compound Interest

In Chap. 4 (Problem 4-6 and its solution), we saw an example of what happens when interest on a certificate of deposit (CD) accumulates over time. *Compound interest* applies not only to the original amount of cash saved (called the *principal*), but also on any interest that has accumulated up to the time you get a new interest payment. You might hear that interest compounds annually (once a year), quarterly (every three months), monthly, weekly, or even daily.

You'll see a figure for the percentage of interest applied each time, and also an equivalent *annual percentage rate* (APR).

TIP *For any particular rate of interest, the APR increases (relative to the rate when compounded annually) as the interest gets paid more often. If somebody tells you that a savings account "pays 3% interest," always ask how often it's compounded! The more often, the better.*

TIP *Obviously, you should always look for the best possible interest rate when you want to put money in a savings account or certificate of deposit. Conversely, if you take out a loan, you should always look for the lowest interest rate. In a perfect world, that's all you'd have to worry about. In real life, you'll also have to concern yourself with the reputation and security of the bank itself. You can obtain expert information about banks and other savings institutions by visiting a Web site called bankrate.com.*

PROBLEM 8-3

Let's revisit the situation described all the way back in Problem 4-6: You buy a five-year CD for \$1000.00, and it earns interest at 2% per year. The bank rounds the value off to the nearest penny at the end of each year, so the interest compounds annually. After the first year, you can expect to have \$1000.00 × 1.02 = \$1020.00 in the account. Now suppose that the bank decides to compound your interest quarterly rather than annually. To compensate for the more frequent payments, instead of giving you 2% after one year, they give you 1/4 that much, or 0.5%, every three months. If you look at the scenario in a simplistic fashion, it might seem to give you the same 2% that you'd have gotten in the original case. But actually, you'll receive a little more. How much more will you have at the end of a year when you get paid 0.5% quarterly, as compared with the situation where you get paid 2% just once at the end of the year?

SOLUTION

In order to work out this problem, you should multiply \$1000 by 1.005 (that's 100.5% of \$1000, representing an increase of 0.5%) at the end of three months, then multiply that amount by 1.005 again after three more months, and repeat the process twice more, rounding off to the nearest penny each time. Here's what happens:

- After three months: \$1000.00 × 1.005 = \$1005.00
- After six months: \$1005.00 × 1.005 = \$1010.03

- After nine months: $1010.03 × 1.005 = $1015.08
- After one year: $1015.08 × 1.005 = $1020.16

You'll have 16 cents more. That's not much, but if you have a lot of money and keep it in savings for a long time, or if the interest rate is better, the situation will differ drastically.

PROBLEM 8-4

Suppose that you find a better deal than your bank has to offer. You invest in a company that ends up paying you the equivalent of 12% interest on that original $1000.00 investment. If they pay you annually, you'll have $1000.00 × 1.12 = $1120.00 at the end of the year. But in fact, they decide to pay you quarterly, giving you 3% (1/4 of 12%) at the end of each three-month period. How much difference will the quarterly payment scheme make compared with the annual payment scheme in this situation?

SOLUTION

You should multiply $1000.00 by 1.03 (that's 103% of $1000, representing an increase of 3%) at the end of three months, then multiply that amount by 1.03 again after three more months, and repeat the process twice more, rounding off to the nearest penny each time. Your investment grows as follows:

- After three months: $1000.00 × 1.03 = $1030.00
- After six months: $1030.00 × 1.03 = $1060.90
- After nine months: $1060.90 × 1.03 = $1092.73
- After one year: $1092.73 × 1.03 = $1125.51

You'll have $5.51 more with the four quarterly payments than you'd get with the single annual payment. Not only does a higher interest rate itself give you more "on its face," but if you can arrange to have it compounded more often, your advantage over the annualized rate increases as well.

PROBLEM 8-5

Suppose that you have acquired great skill as a negotiator, and you get the company described above to pay your interest compounded monthly. You get 1% at the end of each month 12 times throughout the year. How does this scheme compare with the single annual interest payment of 12%? How does it compare with the situation where the interest is compounded quarterly?

SOLUTION

You multiply \$1000.00 by 1.01 (that's 101% of \$1000, representing an increase of 1%) at the end of one month, then multiply that amount by 1.01 again after one more month, and repeat the process 10 more times, rounding off to the nearest penny each time. Your investment grows month by month on the following schedule:

- **After one month: \$1000.00 × 1.01 = \$1010.00**
- **After two months: \$1010.00 × 1.01 = \$1020.10**
- **After three months: \$1020.10 × 1.01 = \$1030.30**
- **After four months: \$1030.30 × 1.01 = \$1040.60**
- **After five months: \$1040.60 × 1.01 = \$1051.01**
- **After six months: \$1051.01 × 1.01 = \$1061.52**
- **After seven months: \$1061.52 × 1.01 = \$1072.14**
- **After eight months: \$1072.14 × 1.01 = \$1082.86**
- **After nine months: \$1082.86 × 1.01 = \$1093.69**
- **After 10 months: \$1093.69 × 1.01 = \$1104.63**
- **After 11 months: \$1104.63 × 1.01 = \$1115.68**
- **After one year: \$1115.68 × 1.01 = \$1126.84**

You'll have \$1.33 more with the 12 monthly payments than you'd get with the four quarterly payments (\$1126.84 − 1125.51), and \$6.84 more than you'd get with the single annual payment (\$1126.84 − \$1120.00).

Still Struggling

Generally speaking, the more money you invest, the better the interest rates you can expect to receive, and the more often you can hope to get your interest compounded. More cash on hand means better deals for you! That's how "the rich get richer"—assuming that they invest their money wisely and don't fall victim to bad luck. Imagine what could happen if, instead of \$1000.00, you had \$10 million to invest, like some major corporate executives have! All of the aforementioned amounts and advantages would increase by a factor of 10,000 (or, if you prefer to use more technical jargon, four orders of magnitude).

Loans and Amortization

When you take out a loan with interest, you pay for the privilege of using someone else's money. As the interest rate increases, you pay more per dollar of money borrowed; most people intuitively know that. Fewer people realize (or take seriously) the fact that, for a given fixed rate of interest, you pay more per dollar of money borrowed as you take longer to pay the loan off. Accountants use the term *amortization* when they talk about the process or schedule of paying off a loan.

If you take out a loan for a period of years, the lender will usually calculate the total interest that you'll owe over the period of the entire loan. Then, in the early months and years, the majority of each payment that you make will go towards that total interest. The actual amount of money you borrow, called the *principal*, gets paid off more and more rapidly as the years go by. You can see how this process works in detail by looking at so-called *amortization tables*.

You can find amortization calculators on the Internet that allow you to compare what happens with repayment schedules for a loan with a fixed principal, but variable interest rates and amortization periods. Enter the phrase "amortization calculator" into your favorite search engine. You'll get a selection of programs that you can experiment with. The following three problems provide examples of what you can expect as you "learn by doing." Do you plan to buy a house and take out a loan (a *mortgage*) to buy it? Do you plan to make an initial cash payment (a *down payment*) to reduce the amount of money that you'll actually have to borrow? How much money, per month, can you afford to pay? Will the loan include a large lump sum payment due at the end of the term (a *balloon payment*)?

PROBLEM 8-6

Suppose that the interest rate on a loan doesn't change, and the amount of the loan doesn't change either. However, you speed up the amortization process by making larger individual payments because you want to get out of debt sooner. What happens to the total amount of interest that you have to pay over the entire course of the loan, as a function of how fast you pay it off?

SOLUTION

The total amount of interest goes down as the amortization goes faster. That's the advantage of having a short-term amortization (say, 10 or

15 years) instead of a longer amortization period (say, 30 years). Go to one of the amortization calculator Web sites and compare the total amounts of interest that you would have to pay on a $150,000 loan at 5.5% amortized over 10 years, 15 years, and 30 years.

PROBLEM 8-7

If the interest rate on a loan doesn't change and the amount of the loan doesn't change but the amortization goes faster, what happens to the monthly payment amounts?

SOLUTION

The monthly payments increase as the amortization goes faster. That's the main disadvantage of having a short-term loan rather than a long-term one. Go to one of the amortization calculator Web sites and compare the monthly payments that you'll have to make on a $150,000 loan at 5.5% amortized over 10 years, 15 years, and 30 years.

PROBLEM 8-8

If the amount of a loan and the amortization period both remain constant, what happens as the interest rate increases?

SOLUTION

The monthly payments and the total interest that you pay both increase. Go to one of the amortization calculator Web sites and compare the monthly, annual, and total payments that you'll have to make on a $150,000 loan at 5.5%, 6.0%, 6.5%, and 7.0%, amortized over 30 years. Also check out the amount of interest that you ultimately have to pay, compared to the principal ($150,000).

TIP *You should always try to negotiate the lowest possible interest rate for a loan. Over the course of many years, a seemingly small difference in the interest rate can make a huge difference in the total amount of money you have to pay to the lender.*

Still Struggling

A good amortization calculator allows you to explore all of your options when you plan to take out a long-term loan. Some of those programs will show your payment schedule by the year. A few of them will show you the schedule, down to the penny, for long-term loans (up to 30 years) on a month-by-month basis.

Taxation

Taxes comprise money that people pay to a government in exchange for services. At the local level, such services include police protection, fire departments, water and sewer services, trash collection, and street maintenance. At the state or provincial level, we get social services of various sorts, specific medical services, welfare programs, highway maintenance, and business regulation, to name a few. At the national level we have military defense, medical care for the poor, and programs for the elderly (Social Security and Medicare in the United States, for example).

All taxes, no matter at what level or for what purpose, can be broadly classified into two major categories: *progressive* and *regressive*.

A progressive tax goes up *disproportionately* as your ability to pay goes up. That's how the *income tax* system works in the United States. The government operates on the principle that rich people can afford to pay a greater *percentage* of their income than poor people can (not just greater *amounts*). In fact, the poorest people might get money from the government in the form of payments called *negative income tax*.

A regressive tax places a greater *effective* burden on poor people than it imposes on rich people. Even if the actual percentage or rate is the same for everybody, a regressive tax "hurts" people more as they earn *less* money. The most common examples of regressive taxes are the *retail sales tax* and its "cousin" the *value-added tax*, especially when they apply to essential goods and services, such as food, housing, and medical care.

PROBLEM 8-9

In recent years, some people in the United States have called for the implementation of a *flat income tax* to replace the current progressive or "bracketed"

income tax system. With a flat tax scheme, everyone would pay the same percentage of their income (the so-called *flat rate*). Critics call such a system regressive. Why?

SOLUTION

Suppose that a flat income tax is imposed at the rate of 20%. If you earn \$20,000 in the year 2014 and you want to feed, clothe, and house a family of four, you'll have a difficult time getting by, even if you don't get taxed at all. If the government takes 20% of that \$20,000, that means you lose $0.20 \times \$20,000 = \$4000$ before you can spend a single penny. You have only \$16,000 left then, and you'll need a lot of luck to live on that!

If you earn \$200,000 in the year 2014 for a family of four and the government takes 20% of that amount, then the government takes away $0.20 \times \$200,000 = \$40,000$, a much larger sum in absolute terms than they take if you earn only \$20,000 before the tax. But with the higher salary, you have \$160,000 remaining. The aforementioned head of household would gladly pay the 20% tax if her income were to rise overnight from \$20,000 to \$200,000.

TIP *Proponents of a flat tax usually incorporate an exemption for the first \$20,000 or \$40,000 or even \$100,000 earned per year to offset the inherent regressive nature of the tax. In that system, people would pay the stated rate of tax only on income in excess of the exemption threshold. People earning less than the threshold amount wouldn't have to pay any tax at all.*

PROBLEM 8-10

How does a retail sales tax work?

SOLUTION

A retail sales tax gets charged to the final purchaser of a product or service, usually based on the location where the customer makes the purchase. If you buy a shirt that carries a retail price of \$30.00 and the sales tax is 5%, then you must pay an additional 5% of \$30.00, or \$1.50, to the store owner, who must, in turn, send \$1.50 to the government entity (the state, for example) that imposes the tax. You end up actually paying \$31.50 for the shirt. The retailer doesn't lose any money because of the tax, but you do!

As its name implies, a retail sales tax applies only to the final (ultimate) purchase. Wholesalers, manufacturers, or other middle and originating businesses don't have to deal with it. Proponents of the sales tax argue that it provides a more reliable source of revenue than an income tax does because people must pay the tax even if they don't have any income (unless, of course, they don't buy anything either). Critics of the sales tax complain that it's regressive, especially when it applies to necessities, such as food, electricity, heating fuel, and vital services because it imposes a greater proportional financial burden on poor people than it puts on rich people.

PROBLEM 8-11

How does a value-added tax work? How does it differ from a retail sales tax? How does it resemble a retail sales tax?

SOLUTION

A value-added tax (often called by its acronym VAT) is imposed at every level of the process where a certain item or service makes its way from the originator to the final consumer. The tax does not apply to the actual price, but only to the amount by which a given business increases the price—the *value added* by that entity. From the point of view of the buyer, the VAT looks like a sales tax. The government receives the same total amount of money from, say, a 5% VAT as it would get from a 5% retail sales tax, but the revenue comes from multiple sources (all the businesses in the supply chain) rather than a single source (the retailer).

To a business retailer, wholesaler, producer, or other intermediary, the VAT works a lot differently than a retail sales tax does. Every business must deal with the tax, not only the retailer, so the total amount of paperwork is much greater with the VAT than it is with a simple retail sales tax. The government needs to hire extra bureaucrats to administer the VAT and shuffle all those papers (or computer files) around. However, a VAT doesn't generate any more revenue than a retail sales tax does at the same rate. The VAT, like the retail sales tax, operates in a regressive fashion.

TIP *While retail sales taxes usually don't apply to essential goods and services, such as food and medicine, VATs usually do. Poor people can't get away without things like food and medicine; so in practice, a poorly or unwisely administered VAT turns out to be even more regressive than a retail sales tax. Not only that, but*

VAT rates are usually higher than retail sales tax rates. In some European countries, for example, the VAT exceeds 20%. Imagine having to pay a tax like that to the federal government on top of your own state's sales tax!

PROBLEM 8-12

Suppose that you have a coop full of chickens who lay the world's best eggs. You sell your eggs to a wholesale distributor for $1.00 a dozen. The distributor sells the eggs to retail stores for $2.00 a dozen. The retail stores sell them for $4.00 a dozen. The federal government imposes a 10% VAT. How does each entity in the chain handle the VAT collection and payment process? Assume that the VAT laws allow no deductions or exemptions whatsoever. Also assume that the VAT is imposed and calculated independently of all other taxes.

SOLUTION

Let's break the process down into each of its several steps. First, you have to feed and care for the chickens, but the VAT doesn't allow you to deduct anything for the cost of all that work and material. You have to pay the government 10% of all the money you get for your eggs. When you sell the wholesaler a dozen eggs, you have to send $0.10 \times \$1.00$, or 10 cents, to the government. Then the following sequence of events unfolds.

- You charge an extra 10 cents to the wholesaler, so she has to pay you $1.10 for those dozen eggs. That way, you get to keep the entire dollar to which you're entitled; you don't have to absorb the tax.
- The wholesaler offers those same dozen eggs to a retailer for $2.00. The value added is $2.00 (the net wholesale price) minus $1.00 (your net price) or $1.00, so the wholesaler sends another 10 cents ($0.10 \times \$1.00$) to the government.
- The wholesaler has already paid you 10 cents and has also paid the government 10 cents, so he charges the retailer the extra 20 cents, for an actual price of $2.20. That way, the wholesaler gets to keep the entire $1.00 of value added at his stage of the game.
- The retailer buys the dozen eggs from the wholesaler for $2.20. Then she puts them on the shelf for shoppers to find and buy at a quoted net price of $4.00.
- The mayor of the town buys a dozen of those wonderful eggs, so the value added at the retailer is $4.00 (the net retail price) minus $2.00 (the net wholesale price), which works out to $2.00. The retailer, therefore, sends 20 cents ($0.10 \times \$2.00$) to the government.

- The retailer, having actually paid out $2.20 for those eggs and having had to send the government 20 cents more, charges the mayor $4.40 for the eggs. That way, the retailer gets to keep the entire value added at her stage, which equals $2.00.
- The egg-loving mayor ends up bearing the entire burden of the VAT, which is 40 cents (10 cents from you to the wholesaler, 10 more cents from the wholesaler to the retailer, and 20 more cents from the retailer to the mayor).
- Note that 40 cents equals exactly 10% of the net retail price of $4.00. Therefore, although the VAT operates at multiple points, it always "passes through." The process doesn't result in any "tax on a tax" phenomenon (called *tax pyramiding*), which can sometimes occur with poorly administered sales and business taxes.
- The government gets its 10% tax, all of which the consumer actually pays.
- From the consumer's point of view, the VAT looks like a sales tax.

As you can see, the VAT works much differently than a sales tax does. In particular, while a typical sales tax *almost never* applies at intermediate levels in a supply chain, a VAT *almost always* does. As you can also see, the VAT is more complicated than a simple retail sales tax.

TIP *To learn more about how a VAT system works, enter the phrase "value-added tax" into your favorite Internet search engine. It has a fascinating history, and its rapid spread bears testimony to governments' fondness for it throughout the world. (As for whether or not working people like the VAT in countries that have it, you'll have to ask them.)*

Still Struggling

At this point you'll probably wonder, "Why would a government impose a VAT when they could get as much money, with a lot less bureaucracy, from a simple retail sales tax?" The answer lies in the fact that a VAT is easier to conceal from the consumer than a sales tax is because the VAT usually doesn't show up on a sales receipt, whereas the sales tax usually does. Its hidden nature allows governments to impose it at a higher rate than people would likely accept for a retail sales tax. In addition, the VAT is more difficult to evade than a sales tax is because each entity in the process must pay something to the government. Rather than absorbing the VAT, each business charges its customer or client the amount of the tax it had to pay, so the VAT has a built-in self-enforcement mechanism. Everybody except the retail consumer ends up doing some paperwork, but a VAT gets better compliance than a retail sales tax does. Governments, especially in Europe, have grown fond of the VAT because it, in addition to the income tax, has proven to be an excellent source of revenue for programs and services that the public demands. Critics say that it encourages excessive government spending, and they also dislike its inherent regressive nature.

PROBLEM 8-13

How does an *estate tax* differ from an *inheritance tax*?

SOLUTION

Both of these tax types (called "death taxes" by their detractors) were originally designed to ensure that rich families would not evolve into dynasties, although neither tax has quite lived up to that expectation. True dynasties have a way of surviving every humanmade adversity. The difference lies in the mode of tax administration. With an estate tax, the money goes to the government *before* any of the inheritance benefits go to the heirs. With the inheritance tax, the heirs pay the tax *after* they receive the benefits.

While the inheritance-tax scheme might at first seem better for the heirs than the estate tax does, in practice the opposite usually happens. Inheritances often come in the form of *hard assets,* such as real estate, but the government demands payment of the tax in cash. As a result, the heirs may face a financial crisis on top of the loss of their loved ones. That's the reason why, over the years, inheritance taxes have been replaced by estate taxes in most jurisdictions.

TIP *The federal "death tax" in the United States is technically an estate tax, not an inheritance tax. However, some states still impose true inheritance taxes.*

TIP *The heirs of small businesses and family farms have complained in recent years about inheritance and estate taxes in the United States. When the head of the enterprise dies, the heirs may have to sell part or all of the business or farm to pay the tax.*

Still Struggling

Whenever you hear someone talk about an "inheritance tax," be sure that you know whether he refers to a true inheritance tax or to an estate tax. For some reason, the two terms have gotten hopelessly mixed up. In addition, you should be aware that both tax types usually apply only to estate values in excess of a certain threshold, ranging from about $1 million up to several million.

PROBLEM 8-14

How does the Social Security tax differ from the income tax in the United States?

SOLUTION

The Social Security tax applies from the very first penny of wages, salaries, or net profit; you don't get any deductions for things like charitable contributions. The tax is imposed at a flat rate, but only on amounts *less* than a certain threshold (approximately $100,000 in the United States at the time of this writing). Therefore, the Social Security tax is sharply regressive. The income tax, in contrast, applies only to amounts *more* than a certain threshold, which varies depending on the number and amounts of deductions that you are allowed to take. In addition, the tax rate generally increases as your income increases, so the income tax is progressive.

PROBLEM 8-15

Suppose that you earn $20,000 a year and the government takes 7% in the form of Social Security tax. How much money does that percentage represent, assuming that the "tax ceiling" or "tax cap" is set at $100,000?

SOLUTION

Seven percent of $20,000 equals $1400. Your income falls well below the "cap," so the tax applies to all of it.

PROBLEM 8-16

Now imagine that you earn $200,000 a year and the government takes 7% for Social Security tax, but with the same "cap" of $100,000. How much money do you end up parting with to pay the tax in that case? What percentage of your total income does it represent?

SOLUTION

You pay only 7% of the first $100,000 that you earn, and that works out to $7000. All of your income above the $100,000 "cap" is tax-free. Therefore, your total percentage paid out equals $100 \times (\$7000 / \$200{,}000) = 3.5\%$—*half* the percentage that you'd pay if you earned *one-tenth* as much money!

QUIZ

Refer to the text in this chapter if necessary. A good score is eight correct. Answers are in the back of the book.

1. **If you pay for a $2.50 item with a $5.00 bill, how much change should you get?**
   A. $2.00
   B. $7.50
   C. $2.00
   D. $2.50

2. **How might a competent salesperson *correctly and properly* count back your change in the situation of Question 1?**
   A. Hand you five dimes (10-cent pieces) and say "sixty, seventy, eighty, ninety, three," and then hand you two $1.00 bills and say "four, five dollars."
   B. Hand you three $1.00 bills and say "three, four, five dollars."
   C. Hand you eight quarters (25-cent pieces) and say "seventy-five, six dollars, twenty-five, fifty, seventy-five, seven dollars, twenty-five, and fifty cents."
   D. Hand you two $1.00 bills and say "one, two," and then hand you two quarters and say "twenty-five, fifty cents."

3. **When interest on a savings account compounds more often than once a year, the effective annual percentage rate (APR)**
   A. equals the interest rate that you would get if it were paid only once a year.
   B. exceeds the interest rate that you would get if it were paid only once a year.
   C. is less than the interest rate that you would get if it were paid only once a year.
   D. becomes negative.

4. **You can maximize the total amount of interest that you get on a long-term savings account or a certificate of deposit (CD) by**
   A. increasing the initial amount that you put in.
   B. getting the highest APR that you can find.
   C. keeping your money in the account for as long as possible without withdrawing any.
   D. All of the above

5. **Under what circumstances would you want to *minimize* the APR for interest on a sum of money?**
   A. When you invest it in a CD
   B. When it's a loan that you make to someone else
   C. When it's a loan that a bank makes to you
   D. Never

6. **Suppose that our government enacts a flat income tax system with an exemption for the first $25,000 earned per year. If you make $40,000 in a given year and**

the tax rate is 20%, what percentage of your total income will you actually pay to the government in that year?

A. 5.00%
B. 7.50%
C. 12.25%
D. 13.75%

7. **Imagine the same tax system as the one described in Question 6, but in the next year you find a new job with double the salary—$80,000 instead of $40,000. What percentage of your income will you actually pay to the government in that year?**

   A. 5.00%
   B. 7.50%
   C. 12.25%
   D. 13.75%

8. **Suppose, as in an example described earlier in this chapter, that you own some chickens who lay the world's best eggs. Inflation has made you decide to raise your price. You sell your eggs to a wholesale distributor for $1.50 a dozen. The distributor sells the eggs to retail stores for $2.75 a dozen. The retail stores sell them for $5.00 a dozen. The federal government imposes a 10% VAT. How much money does the distributor actually lose as a result of the VAT, compared with the situation that would exist if there were no tax at all?**

   A. Nothing
   B. 12.5 cents
   C. 22.5 cents
   D. 27.5 cents

9. **In the scenario of Question 8, how much does the consumer at the store actually lose per dozen eggs as a result of the VAT, compared with the situation that would exist if there were no tax at all?**

   A. 15 cents
   B. 22.5 cents
   C. 27.5 cents
   D. 50 cents

10. **In the scenario of Question 8, how much VAT must you, the chicken-egg cultivator, charge the distributor and then remit to the government per dozen eggs?**

    A. 7.5 cents
    B. 12.5 cents
    C. 15 cents
    D. 27.5 cents

chapter 9

Extremes

Scientists and engineers use a special trick to make calculations involving gigantic and minuscule numbers. How many grains of sand populate the beaches of the earth? What's the ratio of the diameter of a golf ball to the diameter of the moon? How many times brighter is the sun than a tallow candle? *Scientific notation* lets us express such quantities without having to scribble out long strings of digits.

CHAPTER OBJECTIVES

In this chapter, you will

- Discover how subscripts and superscripts add meaning to symbols.
- Contrast positive and negative powers.
- Learn to do arithmetic with numbers in scientific notation.
- Compare truncation and rounding as methods of approximation.
- Use rules of arithmetic precedence to simplify complicated expressions.
- See how significant figures can define precision and accuracy.

Subscripts and Superscripts

Subscripts modify the meanings of units, constants, and variables. A subscript goes to the right of the main character below the "base line." *Superscripts* usually represent *exponents* (powers, like we use when we square or cube a number). A superscript goes to the right of the main character above the "base line."

Numeric subscripts never appear in italics, but alphabetic subscripts sometimes do. Here are three examples of subscripted quantities:

Z_0 read "*Z* sub nought" or "*Z* sub zero"

R_{out} read "*R* sub out"

y_n read "*y* sub *n*"

Numeric superscripts never appear in italics, but alphabetic superscripts usually do. Examples of superscripted quantities are:

2^3 read "two cubed"

10^x read "10 to the *x*th power"

$10^{1/2}$ read "*y* to the one-half power"

Powers of 10 Revisited

Scientists and engineers express extreme numerical values using *power-of-10 notation*, also called *scientific notation*. We write a numeral in *standard power-of-10 notation* according to the following format:

$$m\,.\,n_1 n_2 n_3 \cdots n_p \times 10^z$$

where the dot (.) goes on the base line and represents a decimal point. The numeral *m* (to the left of the decimal point) can be any of the digits 1, 2, 3, 4, 5, 6, 7, 8, or 9. The numerals n_1, n_2, n_3, and so on up to n_p (to the right of the decimal point) can be any of the digits 0, 1, 2, 3, 4, 5, 6, 7, 8, or 9. The value of *z*, which represents the power of 10, can equal any positive or negative whole number (integer), or 0. Here are three examples of numbers written in scientific notation:

$$2.56 \times 10^6$$

$$8.0773 \times 10^{-18}$$

$$1.000 \times 10^0$$

The multiplication sign in a power-of-10 expression can be denoted in various ways. Most scientists in America use the cross symbol (×), as in the above examples. But a small dot raised above the base line (·) is sometimes used to represent multiplication in power-of-10 notation. When written that way, the above numbers look like this in the standard form:

$$2.56 \cdot 10^{6}$$

$$8.0773 \cdot 10^{-18}$$

$$1.000 \cdot 10^{0}$$

TIP ***Don't confuse the small elevated dots in the foregoing expressions with decimal points. That's an easy mistake to make, and a good argument for not using the raised dot to indicate multiplication!***

Still Struggling

If the negative exponent in the middle expression above confuses you, remember what you learned about them in Chap. 3. A negative-integer power is a nonzero quantity divided by itself more than once. The quantity 10^{-18} means 1 divided by 10^{18}, which you would write out in full as

$$1 / 1{,}000{,}000{,}000{,}000{,}000{,}000$$

That's *one quintillionth*, a tiny number indeed!

Another alternative multiplication symbol in power-of-10 notation is the asterisk (*). You'll occasionally see quantities written like this:

$$2.56 * 10^{6}$$

$$8.0773 * 10^{-18}$$

$$1.000 * 10^{0}$$

Once in awhile, you'll have to express numbers in power-of-10 notation using plain, unformatted text. This situation arises, for example, when you must transmit information within the body of an e-mail message (rather than as an attachment). Some calculators and computers use this plain-text scheme. The letter E indicates that the quantity immediately following represents a power of 10. In this format, the above quantities come out as follows:

2.56E6

8.0773E–18

1.000E0

Another alternative involves the use of an asterisk to indicate multiplication, and the symbol ^ to indicate a superscript, so the expressions look like this:

2.56 * 10^6

8.0773 * 10^–18

1.000 * 10^0

Still Struggling

In all of the foregoing examples, the numerical values represented are the same. If we write them out fully in ordinary decimal form, they turn out as follows:

2,560,000

0.000000000000000080773

1.000

As you can see, power-of-10 notation makes it possible to easily write down numbers that denote incredibly big or small quantities. For example, take a look at

$$2.55 \times 10^{45,589}$$

and

$$9.8988 \times 10^{-7,654,321}$$

Imagine trying to write either of these numbers in ordinary decimal form! In the first case, you'd have to put down the numerals 255, and then follow them with a string of 45,587 zeros. In the second case, you'd have to write a numeral zero, then a decimal point, then a string of 7,654,320 zeros, and finally the numerals 9, 8, 9, 8, and 8. Now consider the quantities

$$2.55 \times 10^{45,592}$$

and

$$9.8988 \times 10^{-7,654,318}$$

These numbers look a lot like the first two—but examine them more closely. Both of these new numbers are 1000 times larger than the original two. You can tell by examining the exponents. Both exponents have grown larger by 3. The number 45,592 is 3 more than 45,589, and the number −7,654,318 is 3 larger than −7,654,321. (Numbers grow larger in the mathematical sense as they become more positive or less negative.) The second pair of numbers is three orders of magnitude larger than the first pair of numbers. They look almost the same here, and they would look essentially identical if they were written out in full decimal form. But they differ as much from each other as a meter differs from a kilometer.

As we learned in Chap. 3, the order-of-magnitude concept makes it possible to construct number lines, charts, and graphs with scales that cover huge spans of values. Figure 9-1 illustrates three examples. Drawing A shows a number line spanning three orders of magnitude, from 1 to 1000. Illustration B shows a number line spanning 10 orders of magnitude, from 10^{-3} to 10^{7}. Illustration C shows a graph whose horizontal scale spans 10 orders of magnitude, from 10^{-3} to 10^{7}, and whose vertical scale extends from 0 to 10 in a uniform fashion.

TIP ***While the 0-to-10 scale lends itself easily to the "mind's eye," we can't define how many orders of magnitude it covers. This difficulty arises because, no matter how many times we cut a nonzero number to 1/10 of its original size, we never get all the way down to 0. In a certain intuitive sense, a linear scale (a scale where all the graduations are the same distance apart) ranging from 0 to any positive or negative value covers "infinitely many" orders of magnitude!***

TIP ***In printed literature, power-of-10 notation is rarely used when the power of 10 is between −3 and 3 inclusive. If the exponent is −4 or smaller, or if it's 4 or larger, values are expressed in power-of-10 notation, as a rule.***

TIP ***Scientists use verbal*** **prefix multipliers** ***to express orders of magnitude. Table 9-1 on page 195 shows the prefix multipliers for factors from 10^{-24} to 10^{24}.***

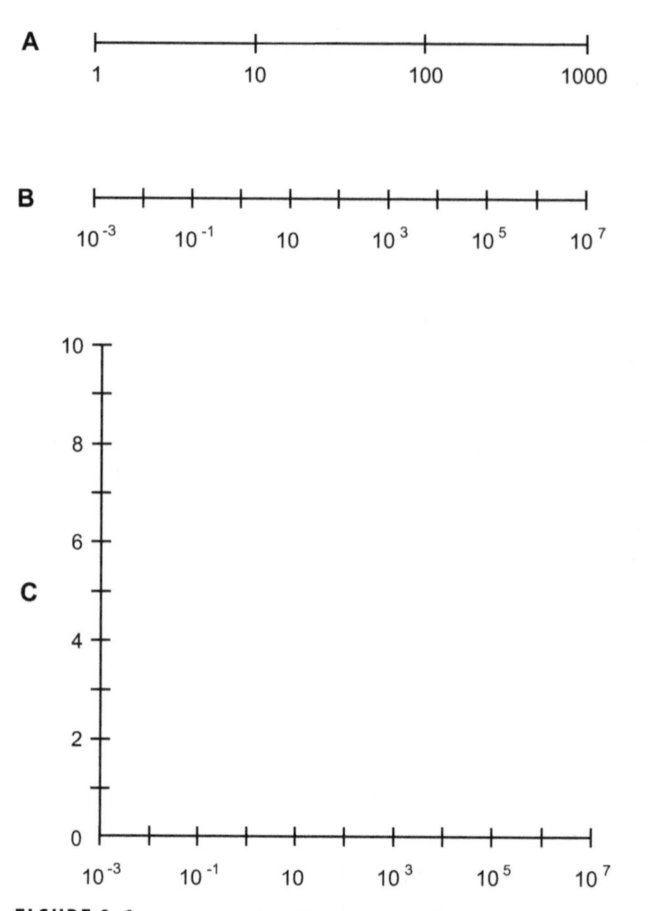

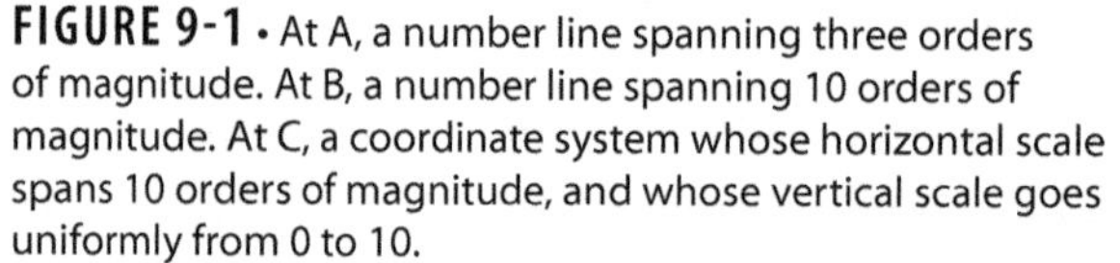

FIGURE 9-1 • At A, a number line spanning three orders of magnitude. At B, a number line spanning 10 orders of magnitude. At C, a coordinate system whose horizontal scale spans 10 orders of magnitude, and whose vertical scale goes uniformly from 0 to 10.

Still Struggling

Some scientific calculators, when set for power-of-10 notation, display all numbers that way, even when it's not necessary. This "uniformity overkill" can cause confusion, especially when the power of 10 equals 0 and the calculator shows a lot of digits. Most people understand the expression 8.407 more easily than 8.407000000E+00, for example, even though those two expressions represent the same quantity.

PROBLEM 9-1

By how many orders of magnitude does a *terahertz* differ from a *megahertz*? (The *hertz* is a unit of frequency, equivalent to a cycle per second.)

SOLUTION

Refer to Table 9-1. A terahertz represents 10^{12} hertz, and a megahertz represents 10^{6} hertz. The exponents differ by 6. Therefore, a terahertz differs from a megahertz by six orders of magnitude.

TABLE 9-1 Prefix multipliers and their abbreviations.

Designator	Symbol	Multiplier
yocto–	y	10^{-24}
zepto–	z	10^{-21}
atto–	a	10^{-18}
femto–	f	10^{-15}
pico–	p	10^{-12}
nano–	n	10^{-9}
micro–	μ or mm	10^{-6}
milli–	m	10^{-3}
centi–	c	10^{-2}
deci–	d	10^{-1}
(none)	—	10^{0}
deka–	da or D	10^{1}
hecto–	h	10^{2}
kilo–	K or k	10^{3}
mega–	M	10^{6}
giga–	G	10^{9}
tera–	T	10^{12}
peta–	P	10^{15}
exa–	E	10^{18}
zetta–	Z	10^{21}
yotta–	Y	10^{24}

PROBLEM 9-2

What, if anything, is wrong with the quantity 344.22×10^7 as an expression in power-of-10 notation?

SOLUTION

It's a legitimate quantity, but it's not in the correct format for scientific notation. The quantity to the left of the multiplication symbol should equal at least 1, but remain smaller than 10. To convert the number to the proper format, we can divide the portion to the left of the "times sign" by 100 to make it 3.4422. Then we can multiply the portion to the right of the "times sign" by 100, increasing the exponent by 2 so that it becomes 10^9. This process gives us the same number, but now it's in the correct power-of-10 format: 3.4422×10^9.

Exponents In Action

When you want to add two numbers in scientific notation, you should convert the numbers to ordinary decimal form before performing the operation. Consider the following three examples.

$$\begin{aligned}
&(3.045 \times 10^5) + (6.853 \times 10^6) \\
&= 304{,}500 + 6{,}853{,}000 \\
&= 7{,}157{,}500 \\
&= 7.1575 \times 10^6
\end{aligned}$$

$$\begin{aligned}
&(3.045 \times 10^{-4}) + (6.853 \times 10^{-7}) \\
&= 0.0003045 + 0.0000006853 \\
&= 0.0003051853 \\
&= 3.051853 \times 10^{-4}
\end{aligned}$$

$$\begin{aligned}
&(3.045 \times 10^5) + (6.853 \times 10^{-7}) \\
&= 304{,}500 + 0.0000006853 \\
&= 304{,}500.0000006853 \\
&= 3.045000000006853 \times 10^5
\end{aligned}$$

Subtraction follows the same basic rules as addition does. You should convert everything to plain decimal form and then subtract. Here's what happens when you subtract the pairs of quantities that you just got done adding.

$$(3.045 \times 10^5) - (6.853 \times 10^6)$$
$$= 304{,}500 - 6{,}853{,}000$$
$$= -6{,}548{,}500$$
$$= -6.548500 \times 10^6$$

$$(3.045 \times 10^{-4}) - (6.853 \times 10^{-7})$$
$$= 0.0003045 - 0.0000006853$$
$$= 0.0003038147$$
$$= 3.038147 \times 10^{-4}$$

$$(3.045 \times 10^5) - (6.853 \times 10^{-7})$$
$$= 304{,}500 - 0.0000006853$$
$$= 304{,}499.9999993147$$
$$= 3.044999999993147 \times 10^5$$

TIP ***When you want to convert a long decimal expression to an expression in scientific notation, you should move the decimal point to the right or left until you have only one nonzero digit to the left of the point (1, 2, 3, 4, 5, 6, 7, 8, or 9). Then count the number of places that you moved the point to the left or right. If you had to move the point to the left (converting 304,499.9999993147 to 3.044999999993147, for example), make the power of 10 equal to the number of places you went (in this case 5). If you had to move the point to the right (converting 0.0003038147 to 3.038147, for example), make the power of 10 equal to the negative of the number of places you went (in this case –4).***

When you want to multiply two quantities by each other in power-of-10 notation, you multiply the *coefficients* (the numbers to the left of the multiplication symbols) together. Then you add the powers of 10. Finally, you simplify the product to the standard form. Here are three examples, using the same three number pairs as before.

$$(3.045 \times 10^5) \times (6.853 \times 10^6)$$
$$= (3.045 \times 6.853) \times 10^{(5+6)}$$
$$= 20.867385 \times 10^{11}$$
$$= 2.0867385 \times 10^{12}$$

$$(3.045 \times 10^{-4}) \times (6.853 \times 10^{-7})$$
$$= (3.045 \times 6.853) \times 10^{[-4+(-7)]}$$
$$= 20.867385 \times 10^{-11}$$
$$= 2.0867385 \times 10^{-10}$$

$$(3.045 \times 10^5) \times (6.853 \times 10^{-7})$$
$$= (3.045 \times 6.853) \times 10^{[5+(-7)]}$$
$$= 20.867385 \times 10^{-2}$$
$$= 2.0867385 \times 10^{-1}$$
$$= 0.20867385$$

You should write out the last number in plain decimal form, not in power-of-10 form because the exponent lies between –3 and 3 inclusive.

TIP ***When you end up with a power-of-10-notation coefficient where the number to the left of the decimal point equals something other than a single nonzero digit (1, 2, 3, 4, 5, 6, 7, 8, or 9), you should move the decimal point to whatever spot will give you a single nonzero digit to its left. Then you should count the number of places you had to move the point. If you moved the point to the left, increase the power of 10 by the number of places you went. If you moved the point to the right, decrease the power of 10 by the number of places you went. In the first and second examples above, we moved the point one place to the left to convert 20.867385 to 2.0867385, so we had to increase the power of 10 by 1 (from 11 to 12 in the first case, and from –11 to –10 in the second case).***

When you want to divide one number by another in power-of-10 notation, you divide the first coefficient (the one in the numerator) by the second coefficient (the one in the denominator). Then you take the power of 10 in the numerator minus the power of 10 in the denominator. Finally, you put the whole thing into standard scientific-notation form. Here's how division works, with the same three number pairs as before.

$$(3.045 \times 10^5)/(6.853 \times 10^6)$$
$$= (3.045/6.853) \times 10^{(5-6)}$$
$$\approx 0.444331 \times 10^{-1}$$
$$\approx 0.0444331$$

$$(3.045 \times 10^{-4})/(6.853 \times 10^{-7})$$
$$= (3.045/6.853) \times 10^{[-4-(-7)]}$$
$$\approx 0.444331 \times 10^3$$
$$\approx 4.44331 \times 10^2$$
$$\approx 444.331$$

$$(3.045 \times 10^5)/(6.853 \times 10^{-7})$$
$$= (3.045/6.853) \times 10^{[5-(-7)]}$$
$$\approx 0.444331 \times 10^{12}$$
$$\approx 4.44331 \times 10^{11}$$

TIP ***Did you notice the wavy equals sign (≈) in the above equations? This symbol means "approximately equals." The numbers here don't divide out neatly, so you must approximate the coefficients in your results.***

When you want to raise a quantity to a power in scientific notation, you raise both the coefficient and the power of 10 to that power, and then assemble the result into a single power-of-10 expression. Here's an example.

$$(4.33 \times 10^5)^3$$
$$= (4.33)^3 \times (10^5)^3$$
$$= 81.182737 \times 10^{(5\times3)}$$
$$= 81.182737 \times 10^{15}$$
$$= 8.1182727 \times 10^{16}$$

TIP ***If you're a mathematics "purist," you'll notice unnecessary parentheses around the number 4.33 in the second line above. I put in those parentheses to minimize the risk of confusion. You'll always do better to ensure that your calculation is***

clearly understood at each step (even if that means putting in a few superfluous symbols) than you'll do if you strive for elegance and forget about the importance of getting the right result!

Let's consider another example, in which the power of 10 is negative, but you raise the whole expression to a positive power.

$$(5.27 \times 10^{-4})^2$$
$$= (5.27)^2 \times (10^{-4})^2$$
$$= 27.7729 \times 10^{(-4\times 2)}$$
$$= 27.7729 \times 10^{-8}$$
$$= 2.77729 \times 10^{-7}$$

To find the root of a number in power-of-10 notation, think of the root as a fractional exponent. The square root is equivalent to the 1/2 power. The cube root is the same as the 1/3 power. In general, the *n*th root of a number (where *n* is a positive integer) is the same as the $1/n$ power. When you think of roots this way, you can multiply quantities out just as you would do with whole-number exponents. Here's an example.

$$(5.27 \times 10^{-4})^{1/2}$$
$$= (5.27)^{1/2} \times (10^{-4})^{1/2}$$
$$\approx 2.2956 \times 10^{[-4\times(1/2)]}$$
$$\approx 2.2956 \times 10^{-2}$$
$$\approx 0.02956$$

Note the wavy equals signs in the second, third, and fourth lines. The square root of 5.27 is irrational , so you must approximate it.

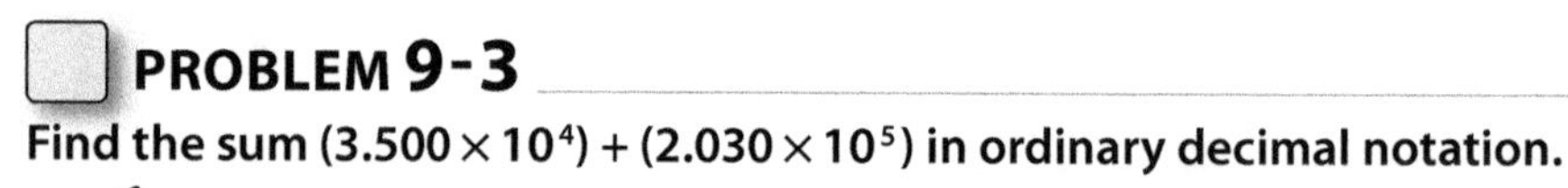

PROBLEM 9-3

Find the sum $(3.500 \times 10^4) + (2.030 \times 10^5)$ in ordinary decimal notation.

SOLUTION

First, expand both of the quantities out so that they appear in plain decimal form as 35,000 and 203,000. Then add to get

$$35{,}000 + 203{,}000 = 238{,}000$$

PROBLEM 9-4

Find the difference $(3.500 \times 10^4) - (2.030 \times 10^5)$ in ordinary decimal notation.

SOLUTION

Take the expanded forms of the numbers and subtract the second one from the first, getting

$$35{,}000 - 203{,}000 = -168{,}000$$

PROBLEM 9-5

Find the product $(3.500 \times 10^4) \times (2.030 \times 10^5)$. Express the answer to three decimal places.

SOLUTION

First, multiply the coefficients to get

$$3.500 \times 2.030 = 7.105$$

Then add the powers of 10 to get

$$10^{(4+5)} = 10^9$$

Finally, assemble the whole expression to obtain the answer 7.105×10^9.

PROBLEM 9-6

Find the quotient $(3.500 \times 10^4)/(2.030 \times 10^5)$. Approximate the answer to four decimal places.

SOLUTION

First, divide the coefficients. You'll get a repeating decimal expression that has a lot of digits in each repetition—a real mess! For now, write out a few more digits than the problem calls for, getting something like

$$3.500 / 2.030 \approx 1.7241379$$

Then subtract the powers of 10 to get

$$10^{(4-5)} = 10^{-1}$$

Next, assemble the expression to obtain

$$1.7241379 \times 10^{-1}$$
$$\approx 0.17241379$$
$$\approx 0.1724$$

trimmed down to four decimal places.

PROBLEM 9-7

Find the quotient $(2.030 \times 10^5)/(3.500 \times 10^4)$. Express the answer to two decimal places.

SOLUTION

First, divide the coefficients to get

$$2.030/3.500 = 0.58$$

Then subtract the powers of 10 to get

$$10^{(5-4)} = 10^1$$

Next, assemble the expression to obtain

$$0.58 \times 10^1 = 5.8$$

PROBLEM 9-8

Simplify the quantity $(2.03 \times 10^5)^4$. Shorten the coefficient to three decimal places.

SOLUTION

First, use your calculator to find the fourth power of 2.03. (You can do that by squaring it, and then squaring that result.) You'll get 16.98181681. Then take 10 to the power of 5×4, which equals 10^{20}. You can now assemble the final result, put it into the correct form, and trim it down like this:

$$16.98181681 \times 10^{20}$$
$$= 1.698181681 \times 10^{21}$$
$$\approx 1.698 \times 10^{21}$$

Approximation and Precedence

Numbers don't always turn out to be exact in the real world. In fact, they almost never do! This fact becomes quite important in observational science and engineering. Usually, we must approximate observed, predicted, or calculated quantities. We can accomplish this task in either of two ways: *truncation* (simple but sometimes inaccurate) and *rounding* (a little tricky, but more accurate).

The process of truncation involves deleting all the numerals to the right of a certain place in the decimal part of an expression. Some older electronic calculators use truncation to fit numbers within their displays. For example, we can truncate the number 3.830175692803 in steps as follows:

$$
\begin{gathered}
3.830175692803 \\
3.83017569280 \\
3.8301756928 \\
3.830175692 \\
3.83017569 \\
3.8301756 \\
3.830175 \\
3.83017 \\
3.8301 \\
3.830 \\
3.83 \\
3.8 \\
3
\end{gathered}
$$

In rounding, when we want to delete a given digit (call it r) at the right-hand extreme of an expression, the digit q to its left (which becomes the new r after the old r is deleted) should stay the same if $0 \le r \le 4$. However, if $5 \le r \le 9$, then we should increase q by 1; in other words, we should *round it up*. Most electronic calculators today use rounding when they have to approximate numbers (for example, when the display doesn't have enough digits to show the entire number perfectly). When we use rounding, the number 3.830175692803 trims down in steps as follows:

$$3.830175692803$$
$$3.83017569280$$
$$3.8301756928$$
$$3.830175693$$
$$3.83017569$$
$$3.8301757$$
$$3.830176$$
$$3.83018$$
$$3.8302$$
$$3.830$$
$$3.83$$
$$3.8$$
$$4$$

Mathematicians agree on a certain order in which operations should be performed when those operations appear together in an expression. This standard prevents confusion and ambiguity. When diverse operations appear in an expression, and if you need to simplify that expression, you should go through the operations in the following sequence:

- Simplify all expressions within parentheses, brackets, and braces from the inside out
- Perform all exponential operations, proceeding from left to right
- Perform all products and quotients, proceeding from left to right
- Perform all sums and differences, proceeding from left to right

Here are two examples of expressions simplified according to the above rules of precedence. Note that the order of the numerals and operations is the same in each case, but the groupings differ.

$$[(2+3)(-3-1)^2]^2$$
$$[5 \times (-4)^2]^2$$
$$(5 \times 16)^2$$
$$80^2$$
$$6400$$

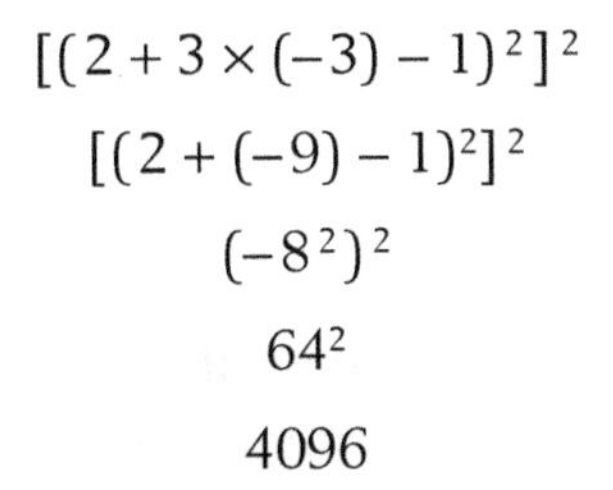

$$[(2 + 3 \times (-3) - 1)^2]^2$$

$$[(2 + (-9) - 1)^2]^2$$

$$(-8^2)^2$$

$$64^2$$

$$4096$$

Now imagine that we have a complicated expression that contains no parentheses, brackets, or braces. We don't have to worry about ambiguity if we strictly adhere to the foregoing set of rules. Consider the following example:

$$z = -3x^3 + 4x^2y - 12xy^2 - 5y^3$$

If we write this expression out with parentheses, brackets, and braces to emphasize the rules of precedence, it comes out as

$$z = [-3(x^3)] + \{4[(x^2)y]\} - \{12[x(y^2)]\} - [5(y^3)]$$

Because we've agreed on the rules of precedence, we can get along without the parentheses, brackets, and braces in this particular case.

PROBLEM 9-9

What's $2 + 3 \times 4 + 5$?

SOLUTION

First, perform the multiplication operation in the middle, obtaining the expression $2 + 12 + 5$. Then add the resulting numbers, obtaining a final value of 19.

PROBLEM 9-10

What's $2 - 3 \times 4 + 2^3$?

SOLUTION

Cube the number 2 first, getting 8, so that the expression becomes $2 - 3 \times 4 + 8$. Then multiply 3 by 4 to get 12, turning the expression into $2 - 12 + 8$. Next, subtract 12 from 2 to get -10, yielding the simple sum $-10 + 8$. Finally, add to get the answer -2.

PROBLEM 9-11

What's 2 − 3 + 4 − 5 + 6?

SOLUTION

In this case, all of the operations have the same precedence, so you should work from left to right straight through the whole thing. First find 2 − 3 = −1. Then add 4 to get 3. Then subtract 5 to get −2. Finally add 6 to get 4.

PROBLEM 9-12

What's 2 − 3 − 4 − 5 − 6?

SOLUTION

As in the previous problem, you should work straight through from left to right because the operations all have the same precedence. First find 2 − 3 = −1. Then subtract 4 to get −5. Then subtract 5 to get −10. Finally subtract 6 to get −16.

Significant Figures

The number of *significant figures*, also called *significant digits*, in a power-of-10 quantity tells us the degree of precision to which we know a particular value, and the extent of accuracy that we can legitimately claim when we perform a calculation.

Whenever we do multiplication, division, or exponentiation in power-of-10 notation, the number of significant figures in the result can't "legally" exceed the number of significant figures in the least-exact expression. You might wonder why, in some of the foregoing examples, we came up with answers that have more digits than the any of the numbers in the original problem. In pure mathematics, where all numbers are exact, we don't have to worry about that sort of thing. But in science and engineering, we must pay attention to accuracy, so significant figures matter a lot.

Consider the two numbers $x = 2.453 \times 10^4$ and $y = 7.2 \times 10^7$. In pure mathematics, the following statement makes perfect sense:

$$\begin{aligned} xy &= 2.453 \times 10^4 \times 7.2 \times 10^7 \\ &= 2.453 \times 7.2 \times 10^{11} \\ &= 17.6616 \times 10^{11} \\ &= 1.76616 \times 10^{12} \end{aligned}$$

But if x and y represent measured quantities, as they would in experimental physics for example, the above statement needs qualification; we must pay close attention to how much accuracy we claim.

TIP ***When you see a product or quotient containing a bunch of numbers in scientific notation, count the number of single digits in the coefficients of each number. Then take the smallest number of digits. That's the number of significant figures you can claim in the final answer or solution.***

In the above example, we have four single digits in the coefficient of x, and two single digits in the coefficient of y. We must, therefore, round off the answer, which appears to contain six significant figures, to only two digits in total. (We should always use rounding, never truncation.) We'll conclude that

$$
\begin{aligned}
xy &= 2.453 \times 10^4 \times 7.2 \times 10^7 \\
&= 1.8 \times 10^{12}
\end{aligned}
$$

Still Struggling

In situations of this sort, if you insist on absolute rigor, you'll use wavy equal signs throughout because you always deal with approximate values. But most folks use ordinary equals signs. Physical quantities, measurements, and calculations never come out exact, and writing wavy lines can get tiresome.

Now suppose that we want to find the quotient x/y instead of the product xy. We proceed as follows:

$$
\begin{aligned}
x/y &= (2.453 \times 10^4)/(7.2 \times 10^7) \\
&= (2.453/7.2) \times 10^{-3} \\
&0.3406944444 \cdots \times 10^{-3} \\
&3.406944444 \cdots \times 10^{-4} \\
&= 3.4 \times 10^{-4}
\end{aligned}
$$

Sometimes, when you make a calculation, you'll get an answer that lands on a neat, seemingly whole-number value. Consider $x = 1.41421$ and $y = 1.41422$.

Both of these have six significant figures. When you multiply this expression out and take significant figures into account, you get

$$\begin{aligned} xy &= 1.41421 \times 1.41422 \\ &= 2.0000040662 \\ &= 2.00000 \end{aligned}$$

This quantity appears to equal exactly 2. In pure mathematics, 2.00000 does indeed mean the very same thing as 2. But in science or engineering, those five zeros indicate how near to the exact number 2 we believe (or can allow ourselves to believe) that the result lies. We know the answer comes very close to a pure mathematician's idea of the number 2, but in the imperfect physical universe, we must say that an uncertainty of up to plus-or-minus 0.000005 (written ± 0.000005) exists. When we claim a certain number of significant figures, the digit 0 carries as much importance as any of the other nine digits in the decimal number system.

TIP ***When you add or subtract measured or estimated quantities, determining the appropriate number of significant figures can involve subjective judgment. You should expand all the values out to their plain decimal form (if possible), make the calculation as if you were a pure mathematician, and then, at the end of the process, decide how many significant figures you can reasonably claim.***

In some cases of addition and subtraction, the outcome of determining significant figures resembles what happens with multiplication or division. Consider, for example, the sum $x + y$, where $x = 3.778800 \times 10^{-6}$ and $y = 9.22 \times 10^{-7}$. The arithmetic calculation proceeds as follows:

$$\begin{aligned} x &= 0.000003778800 \\ y &= 0.000000922 \\ x + y &= 0.0000047008 \\ &= 4.7008 \times 10^{-6} \\ &= 4.70 \times 10^{-6} \end{aligned}$$

In other instances, one of the values in a sum or difference greatly exceeds the other, so that one of them vanishes into insignificance when we take significant figures into account. Let's say that $x = 3.778800 \times 10^4$ and $y = 9.22 \times 10^{-7}$. The process of finding the sum goes like this:

$$x = 37{,}788.00$$

$$y = 0.000000922$$

$$x + y = 37{,}788.000000922$$

$$= 3.7788000000922 \times 10^4$$

In this case, y is so much smaller than x that the value of y doesn't significantly affect the value of the sum. Here, we might as well regard y, in relation to x or to the sum $x + y$, as the equivalent of a dust speck compared with an elephant. If a dust speck lands on an elephant, the total weight does not appreciably change, nor does the presence or absence of the dust speck have any effect on the accuracy of the scales when we weigh the elephant. We can conclude that the "sum" simply equals the larger number:

$$x + y = 3.778800 \times 10^4$$

TIP ***When you have a sum or difference and one value exceeds the other to the extent that you can ignore the smaller one, you can "legally" claim all of the significant figures that the larger number contains. The number of significant figures in the smaller quantity doesn't matter because it might as well not exist!***

PROBLEM 9-13

What's the product 1.001×10^5 and 9.9×10^{-6}, taking significant figures into account?

SOLUTION

You should deal with the coefficients and the powers of 10 separately. When you carry out that process step-by-step, you get

$$(1.001 \times 10^5) \times (9.9 \times 10^{-6})$$

$$= (1.001 \times 9.9) \times (10^5 \times 10^{-6})$$

$$= 9.9099 \times 10^{[5+(-6)]}$$

$$= 9.9099 \times 10^{-1}$$

$$= 0.99099$$

You must round this quantity off to two significant figures because that's the most accuracy that you can "legally" claim. You don't have to write the expression out in scientific form because the power of 10 lies within the range −3 to 3 inclusive. Therefore, you can say that

$$(1.001 \times 10^{5}) \times (9.9 \times 10^{-6}) = 0.99$$

PROBLEM 9-14

What's the quotient 1.001×10^5 divided by 9.9×10^{-6}, taking significant figures into account?

SOLUTION

As in the solution to Problem 9-13, you should work with the coefficients and the powers of 10 separately. As you do the arithmetic and then round the result to two significant figures, you obtain

$$(1.001 \times 10^{5}) / (9.9 \times 10^{-6})$$
$$= (1.001 / 9.9) \times (10^{5} / 10^{-6})$$
$$= 0.101111111 \cdots \times 10^{[5-(-6)]}$$
$$= 0.101111111 \cdots \times 10^{11}$$
$$= 1.001111111 \cdots \times 10^{10}$$
$$= 1.0 \times 10^{10}$$

Still Struggling

Do you wonder about all those extra digits in the intermediate steps of the preceding calculations? The extra digits in the calculation process will eliminate the danger that you'll end up with a so-called *cumulative rounding error*. In situations where repeated rounding introduces many small errors, the final calculation can come out a full digit or two off, even after rounding to the appropriate number of significant figures at every step in the process. Don't let this potential pitfall snare you!

QUIZ

Refer to the text in this chapter if necessary. A good score is eight correct. Answers are in the back of the book.

1. **When we *truncate* the number 13.578793 to *four significant figures*, we get**
   A. 13.5787.
   B. 13.5788.
   C. 13.57.
   D. 13.58.

2. **When we *round* the number 13.578793 to *four decimal places*, we get**
   A. 13.5787.
   B. 13.5788.
   C. 13.57.
   D. 13.58.

3. **When we simplify the expression $3 \times 4 + 5 \times 6$, what do we get?**
   A. 102
   B. 162
   C. 42
   D. We can't simplify it any further than the way it appears here.

4. **When we simplify the expression $3 + 4 \times 5 + 6$, what do we get?**
   A. 29
   B. 41
   C. 47
   D. We can't simplify it any further than the way it appears here.

5. **What's the product $(2.567 \times 10^5) \times (1.7 \times 10^5)$, taking significant figures into account?**
   A. 4.4×10^{10}
   B. 4.36×10^{10}
   C. 4.364×10^{10}
   D. 4.3639×10^{10}

6. **What's the quotient $(2.567 \times 10^5) / (1.7 \times 10^5)$, taking significant figures into account?**
   A. 1.510
   B. 1.51
   C. 1.5
   D. 2

7. What's the value of $(2.567 \times 10^5)^3$, taking significant figures into account?

A. 1.7×10^9
B. 1.7×10^{16}
C. 1.692×10^9
D. 1.692×10^{16}

8. What's the value of $(2.567 \times 10^5)^{-3}$, taking significant figures into account? Note that the final exponent in this expression equals −3.

A. 5.912×10^{-4}
B. 5.912×10^{-17}
C. 5.9×10^{-4}
D. 5.9×10^{-17}

9. What's the sum $(3.7887 \times 10^{78}) + (4.988 \times 10^{25}) + 366 + 102$, taking significant figures into account?

A. 3.7787×10^{78}
B. 3.778×10^{78}
C. 3.78×10^{78}
D. 3.8×10^{78}

10. What's the sum $(3.7887 \times 10^{-78}) + (4.988 \times 10^{-25}) + 366 + 102$, taking significant figures into account? Note that the powers of 10 in the first two terms are negative.

A. 468.00
B. 468.0
C. 468
D. We can't define it!

chapter 10

Measurements

Physical units lend meaning to numbers. You can say that an object is *this* wide or *that* heavy, something lasts for *this* long or gets *that* hot, or a battery produces *this* much voltage or *that* much current. You might at first think that this chapter is too technical for the average person. But wouldn't you like to have an occasional conversation with a scientist and understand her?

CHAPTER OBJECTIVES

In this chapter, you will

- Summarize the three most common unit systems worldwide.
- Contrast base units with derived units.
- See how prefix multipliers can modify physical units.
- Compare relative quantities by adding or changing prefix multipliers.
- Learn how to convert units from one system to another.

Unit Systems

Technical people have devised various systems of physical units, but only three of them prevail today. The *meter/kilogram/second (mks) system*, also called the *metric system* or the *International System*, finds favor in most of the world. Less often, you'll encounter the *centimeter/gram/second (cgs) system*. The *foot/pound/second (fps) system*, also called the *English system*, remains popular among lay people in the United States and to some extent in Great Britain. Each system has several fundamental, or *base*, units from which all the others arise.

Scientists abbreviate the International System as SI (for the words *Système International* in French). It comprises seven fundamental concepts called *base units*:

- The *meter*, which quantifies *distance* (or *displacement*, if we know the direction)
- The *kilogram*, which quantifies *mass* (proportional to *weight* on the earth's surface)
- The *second*, which quantifies *time* (in terms of certain atoms' behavior)
- The *kelvin*, which quantifies *temperature* (compared to *absolute zero*, the total absence of thermal energy or "heat")
- The *ampere*, which quantifies *electric current* (the movement of charged particles such as *electrons*)
- The *candela*, which quantifies *luminous intensity* (the brightness of a light source)
- The *mole*, which quantifies the *amount of substance* (in terms of the number of *elementary particles* such as *atoms*)

In the centimeter/gram/second (cgs) system, the base units are:

- The *centimeter* (equal to 0.01 meter)
- The *gram* (equal to 0.001 kilogram)
- The second (the same as in SI)
- The *degree Celsius* (where water freezes at 0 degrees and boils at +100 degrees on earth at sea level)
- The ampere (the same as in SI)
- The candela (the same as in SI)
- The mole (the same as in SI)

In the English or fps system, the base units are:

- The *foot* (approximately 30.5 centimeters)
- The *pound* (about 454 grams at the earth's surface)
- The second (the same as in SI)
- The *degree Fahrenheit* (where water freezes at +32 degrees and boils at +212 degrees on earth at sea level)
- The ampere (the same as in SI)
- The candela (the same as in SI)
- The mole (the same as in SI)

Primary Base Units

The fundamental unit of distance is the meter, symbolized by the nonitalicized, lowercase English letter m. The original definition of the meter was based on the notion of dividing the shortest possible surface route through Paris, France between the earth's north pole and equator, into 10 million (10^7) identical units, ignoring surface irregularities, such as hills and valleys, as shown in Fig. 10-1. The circumference of the earth, therefore, worked out to about

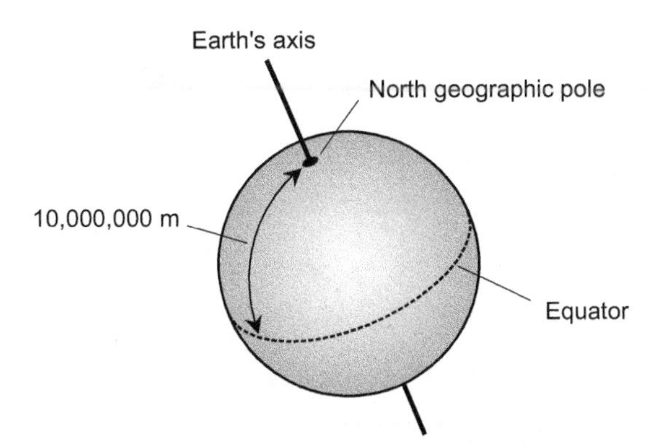

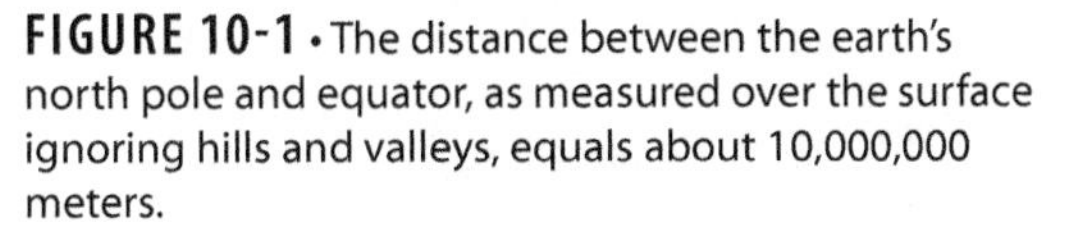
FIGURE 10-1 • The distance between the earth's north pole and equator, as measured over the surface ignoring hills and valleys, equals about 10,000,000 meters.

40,000,000 m. Nowadays, scientists define the meter as the distance that a beam of light travels through a perfect vacuum in 1/299,792,458 of a second. That's approximately the length of an adult's full stride when walking at a brisk pace.

TIP ***Various units smaller or larger than the meter are often employed. A*** **millimeter** ***(mm) equals 0.001 m. A*** **micrometer** ***(μm or μ) equals 0.000001 m or*** 10^{-6} ***m. Some people call a micrometer a*** **micron*****, but that term has technically gone out of date. A*** **nanometer** ***(nm) equals 0.000000001 m or*** 10^{-9} ***m. A*** **kilometer** ***(km) equals 1000 m.***

The base SI unit of mass is the kilogram, symbolized by the lowercase, non-italicized pair of English letters kg. Originally, the kilogram was defined as the mass of a cube of pure liquid water measuring exactly 0.1 m on each edge (Fig. 10-2), equivalent to 0.001 cubic meter. That's still an excellent definition, but these days, scientists have come up with something more absolute. A kilogram represents the mass of a sample of platinum-iridium alloy kept at the *International Bureau of Weights and Measures* in France. This object is called the *international prototype of the kilogram*.

TIP ***Various units other than the kilogram are used for mass. A*** **gram** ***(g) equals 0.001 kg. A*** **milligram** ***(mg) equals 0.001 g. A*** **microgram** ***(μg) equals 0.000001 g or*** 10^{-6} ***g. A*** **nanogram** ***(ng) equals 0.000000001 g or*** 10^{-9} ***g. A*** **metric ton** ***equals 1000 kg.***

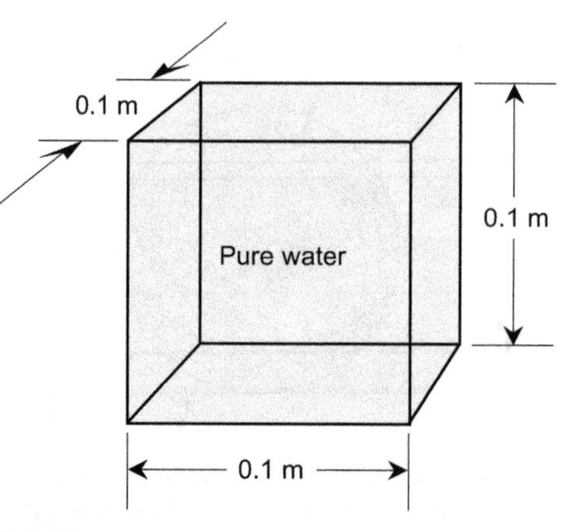

FIGURE 10-2 • Originally, scientists defined the kilogram as the mass of 0.001 cubic meter of pure liquid water.

Still Struggling

Mass varies in direct proportion to weight in any constant gravitational field (such as on the earth's surface), but in technical terms, mass and weight differ. An object with a mass of 1 kg always has a mass of 1 kg, no matter where we take it. That standard platinum-iridium ingot would mass 1 kg on the moon, on Mars, or in interplanetary space. Weight, in contrast, quantifies the force exerted by gravity on an object having a particular mass. On the surface of the earth, an object of mass 1 kg has a weight of about 2.2 pounds.

The SI unit of time is the second, symbolized by the lowercase, nonitalic English letter s. Some people abbreviate it as sec. It was originally defined as exactly 1/60 of a minute, which equals 1/60 of an hour, which equals 1/24 of a *mean solar day*. (Astronomers define the mean solar day as the length of time it takes for the sun to pass from due south to due south again, as seen from any fixed spot in the earth's northern hemisphere and averaged over the course of a whole year.) A second, therefore, constituted 1/60 of 1/60 of 1/24 (or 1/86,400) of a mean solar day, as shown in Fig. 10-3, and that's still a good definition. But more recently, with the availability of atomic time standards, 1 s has

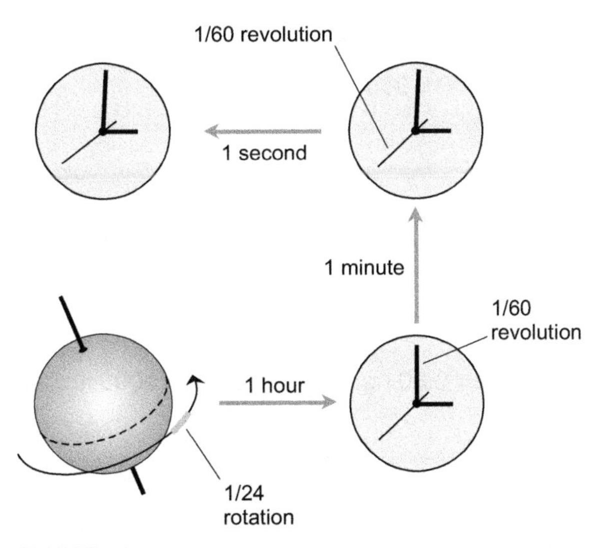

FIGURE 10-3 • Originally, scientists defined the second as the amount of time that passes in 1/60 of 1/60 of 1/24, or 1/86,400, of a mean solar day.

been formally defined as the amount of time it takes for a certain isotope of *cesium 133* (a radioactive element) to oscillate through 9,192,631,770 cycles.

TIP ***Units smaller than the second are often employed to measure or define time. A*** **millisecond** ***(ms) equals 0.001 s; a*** **microsecond** ***(μs) equals 0.000001 s or 10^{-6} s; a*** **nanosecond** ***(ns) equals 0.000000001 s or 10^{-9} s.***

Did You Know?

A second represents roughly the time it takes for a ray of light to travel three-quarters of the way to the moon. You might have heard that the moon orbits at a distance of a little more than one *light-second* from the earth. If you're old enough to remember the conversations that earth-based scientists carried on with the Apollo astronauts as they romped around on the moon, you'll recall the delay between questions from the earthlings and replies from the moonwalkers. It took more than two seconds for the radio signals to make a round trip between earth and the moon!

PROBLEM 10-1

If someone says that a certain object is 0.035 m tall, what's that height in millimeters?

SOLUTION

A distance of 1 mm equals 0.001 m, so 1 m = 1000 mm. Therefore, 0.035 m = 0.035 × 1000 = 35 mm.

PROBLEM 10-2

If someone tells you that a certain object masses 0.035 g, what's its mass in milligrams?

SOLUTION

A mass of 1 mg equals 0.001 g. Therefore, 1 g = 1000 mg, so 0.035 g = 0.035 × 1000 = 35 mg.

PROBLEM 10-3

How many kilometers does a ray of light travel through a vacuum in 1.00 ms? Round off the answer to three significant figures.

SOLUTION

A beam of light travels exactly 1 m through a vacuum in 1/299,792,458 s. It follows that in 1 s, light covers a distance of 299,792,458 m, which equals 299,792.458 km. In 1.00 ms, the same ray goes exactly 0.001 times that far, or 299.792458 km, which we can round off to 300 km.

Secondary Base Units

The SI unit of temperature is the kelvin, symbolized K (uppercase and not italicized). It quantifies the amount of heat that an object has relative to *absolute zero*, which represents the absence of all thermal energy; it's "the coldest possible temperature." A temperature of 0 K represents absolute zero. Kelvin values never get negative. Formally, physicists define the kelvin as a temperature change of 1/273.16, or 0.0036609, of the thermodynamic temperature of the *triple point* of pure water.

TIP ***The triple point equals the temperature and pressure at which pure water can exist as vapor, liquid, and solid in equilibrium. The triple-point temperature is almost, but not exactly, the same as the freezing or melting point.***

The ampere, symbolized by the nonitalic, uppercase English letter A (or abbreviated as amp), quantifies electric current. A flow of approximately 6.24×10^{18} electrons per second, past a given fixed point in an electrical conductor, represents a current of 1 A. The rigorous definition of the ampere is quite arcane! It's the amount of constant *charge-carrier* (such as electron) flow going through two straight, parallel, infinitely thin, perfectly conducting wires, placed exactly 1 m apart in a vacuum, that causes a certain exact amount of force (2×10^{-7} *newton*) to occur per meter of distance along the wires. (We'll learn about the newton shortly.)

TIP ***Various units smaller than the ampere are often used to measure current. One* milliampere *(mA) equals 0.001 A. One* microampere *(μA) equals 0.000001 A or 10^{-6} A. One* nanoampere *(nA) equals 0.000000001 A or 10^{-9} A.***

The candela, symbolized by the nonitalicized, lowercase pair of English letters cd, is the unit of luminous intensity. It represents 1/683 of a *watt* of radiant energy, emitted at a *frequency* of 5.40×10^{14} *hertz* (cycles per second), in a solid angle of 1 *steradian*. (We'll define the watt, the hertz, and the steradian shortly.)

Another definition tells it this way: 1 cd represents the radiation from a surface area of 1.667×10^{-6} square meter of a perfectly-radiating object called a *blackbody*, at the solidification temperature of pure platinum.

TIP *Here's a simple, although crude, definition of the candela: roughly the amount of light emitted by an ordinary candle. Lay people once expressed luminous intensity in units called* **candlepower**. *(A few people still do!) A candlepower is just about the same as a candela. When you hear someone talk about candlepower today, you can be sure that they intend to express the luminous intensity of a light source in candela.*

TIP *Scientists and engineers call the particular thing that a unit quantifies the* **dimension** *of that thing. For example, meters, feet, kilometers, and miles express the distance dimension; seconds, minutes, hours, and days express the time dimension; kelvins, degrees Celsius, and degrees Fahrenheit express the temperature dimension.*

The mole, symbolized or abbreviated by the nonitalicized, lowercase English letters mol, is the standard unit of material quantity. It is also known as *Avogadro's number*, and works out as approximately 6.022×10^{23}. That's the number of atoms in exactly 12 g (0.012 kg) of *carbon 12*, the most common form of elemental carbon with six protons and six neutrons in the nucleus. The mole arises naturally in the physical world, especially in chemistry. It constitutes a *dimensionless unit* because it does not express any particular physical phenomenon. It's merely a standardized number, like a *dozen* (12) or a *gross* (144).

TIP *You'll sometimes hear about units called* **millimoles** *(mmol) and* **kilomoles** *(kmol). A quantity of 1 mmol represents 0.001 mol. A quantity of 1 kmol represents 1000 mol.*

Still Struggling

Until now, we've been rigorous about mentioning that symbols and abbreviations consist of lowercase or uppercase, nonitalicized letters or strings of letters. That's important because if we fail to make that distinction, especially relating to the use of italics, the symbols or abbreviations for physical units can get confused with constants, variables, or coefficients that appear in equations.

When we italicize a letter, it almost always represents a constant, variable, or coefficient. When we don't italicize a letter, it often represents a physical unit or a prefix multiplier. For example, the lowercase letter s (not in italics) stands for "second," but *s* (in italics) often represents displacement. The lowercase letter m (not in italics) represents "meter" or "meters," but *m* (in italics) can denote the slope of a line in a graph.

From now on, let's not belabor this issue every time a unit symbol or abbreviation comes up. But let's not completely forget about it either. As with significant figures in calculations, notational "nits" can make a huge difference in the outcome of a problem or the meaning of an expression!

PROBLEM 10-4

Imagine that you heat a pan of water uniformly at the steady rate of 0.001 K per second from 290 K to 320 K. (Water exists as a liquid at these temperatures on the earth's surface.) Draw a graph of this situation with time on the horizontal axis and temperature on the vertical axis for values between 311 K and 312 K only.

SOLUTION

Figure 10-4 shows the graph. The plot turns out as a straight line, indicating that the temperature of the water rises at a constant rate. Because a change of only 0.001 K takes place every second, you have to wait 1000 seconds for the temperature to rise from 311 K to 312 K. The time scale graduates in relative terms from 0 to 1000.

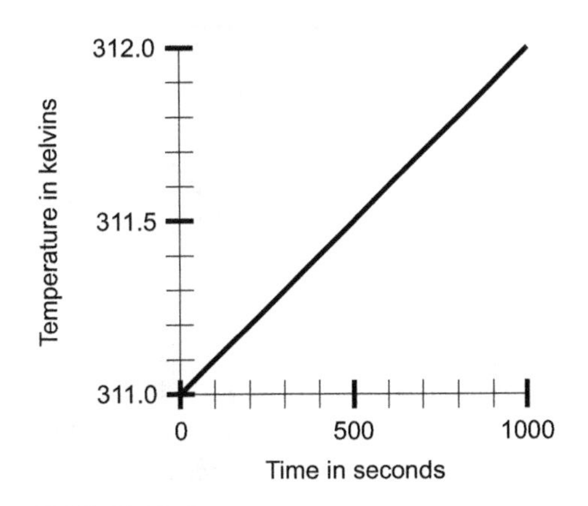

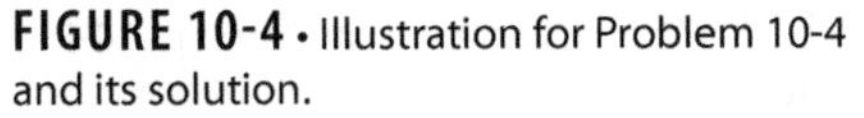

FIGURE 10-4 • Illustration for Problem 10-4 and its solution.

PROBLEM 10-5

Draw a "zoomed-in" version of Fig. 10-4, showing only the temperature range from 311.30 K to 311.40 K. Use a relative time scale, starting at 0.

SOLUTION

Figure 10-5 shows the graph. The time and temperature axes run in the same directions as they do in Fig. 10-4, but we've "magnified" them both by a factor of 10.

PROBLEM 10-6

How many electrons flow past a given point in 3.0 s, given an electrical current of 2.0 A? Express the answer to two significant figures.

SOLUTION

We know that an ampere of current represents the flow of 6.24×10^{18} electrons per second past a point per second of time. Therefore, a current of 2.0 A represents twice this many electrons, or 1.248×10^{19}, flowing past the point each second. (We keep the extra digit in this figure to minimize the danger of ending up with a cumulative rounding error.) In 3.0 seconds, three times that many electrons pass the point. That's 3.744×10^{19} electrons. Rounding back down to two significant figures, we get the answer as 3.7×10^{19} electrons going past the point in 3.0 s when we have a current of 2.0 A.

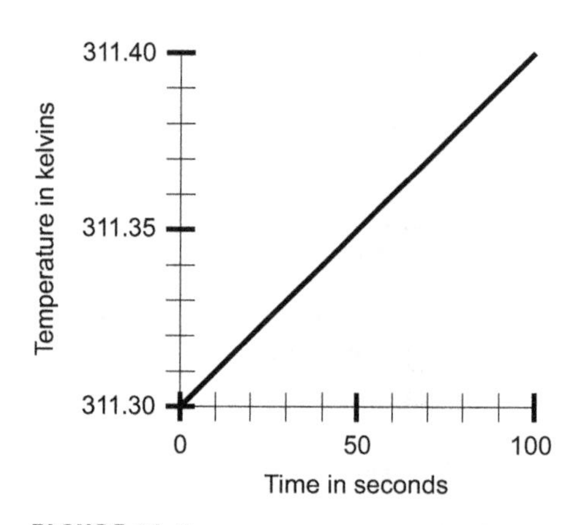

FIGURE 10-5 • Illustration for Problem 10-5 and its solution.

Derived Units

Scientists and engineers combine the base units to generate all sorts of others. Sometimes, the so-called *derived units* are expressed in terms of base units, although such expressions can get confusing (for example, seconds cubed or kilograms to the −1 power). If you ever encounter combinations of units that seem nonsensical, you're witnessing an example of a derived unit that someone put down purely in terms of base units.

The standard unit of *plane angular measure* is the *radian* (rad). It's the angle created by an arc on a circle, whose length, as measured on the circle, equals the radius of the circle. Imagine taking a string and running it out from the center of a circle to some point on the edge, and then laying that string down exactly along the periphery of the circle. The resulting angle at the center will measure 1 rad (Fig. 10-6).

The *angular degree*, symbolized by a little elevated circle (°) or by the three-letter abbreviation deg, equals 1/360 of a complete circle. The history of the degree remains shrouded in uncertainty, although one theory suggests that ancient mathematicians and astronomers chose it because it seemed to be almost exactly the number of days required for the sun to make one complete trip, from west to east, around the group of constellations called the *zodiac*.

The standard unit of *solid angular measure* is the *steradian*, symbolized sr. A solid angle of 1 sr defines a cone with its apex at the center of a sphere, intersecting the surface of the sphere in a circle such that, within the circle, the enclosed area on the sphere equals the square of the sphere's radius (Fig. 10-7).

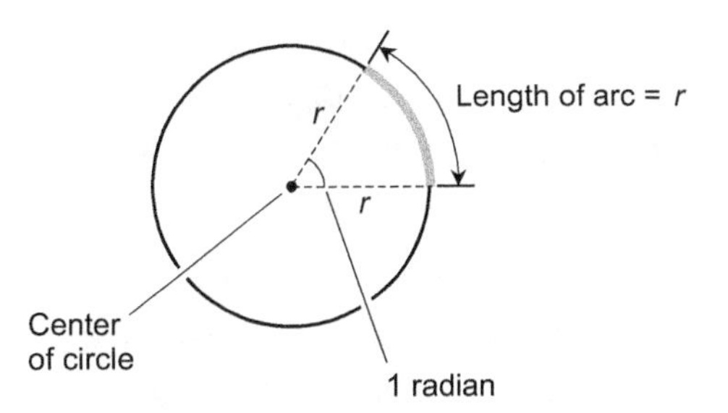

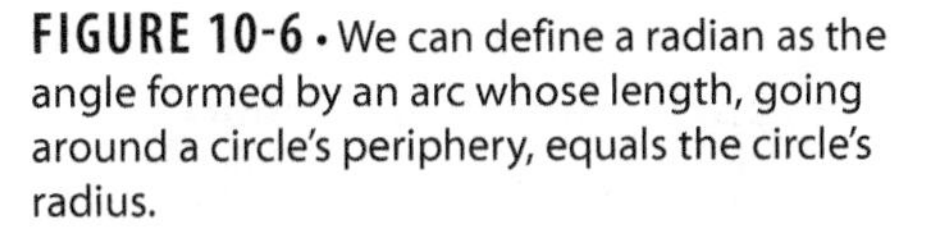

FIGURE 10-6 • We can define a radian as the angle formed by an arc whose length, going around a circle's periphery, equals the circle's radius.

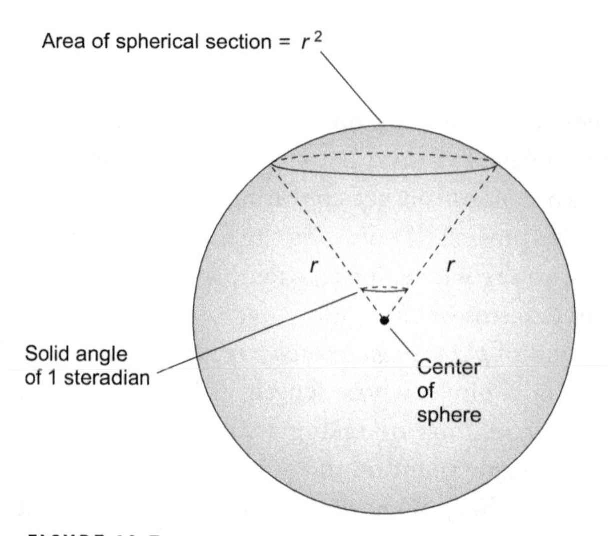

FIGURE 10-7• We can define a steradian as a solid (cone-shaped) three-dimensional angle that cuts through a region on a sphere's surface whose area equals the square of the sphere's radius.

TIP ***One angular degree equals about 0.017453 radians. Conversely, a radian equals roughly 57.296 angular degrees. A complete sphere's surface contains 4π, or approximately 12.566, steradians. Remember that π represents the ratio of a circle's circumference to its diameter, an irrational number equal to about 3.1416.***

Still Struggling

Radians and degrees express angular measures on a flat two-dimensional (2D) surface. Steradians express angular measure in three dimensions (3D). Although their names look and sound similar, the radian and the steradian differ conceptually from each other in the same way as, say, square meters and cubic meters do.

The standard unit of *mechanical force* is the *newton*, symbolized N. One newton represents the force it takes to make a mass of 1 kg accelerate at a rate of one meter per second squared (1 m/s^2) in the absence of other forces such as gravity. Aerospace engineers express jet or rocket propulsion in newtons. Force equals the product of mass and acceleration. Reduced to base units in SI, newtons are equivalent to kilogram-meters per second squared ($kg \cdot m/s^2$).

The standard unit of *energy* is the *joule*, symbolized J. It's a small unit in real-world terms. One joule turns out as the equivalent of a newton-meter ($N \cdot m$). If we reduce the joule to base units in SI, it works out as mass times unit distance squared divided by time squared ($kg \cdot m^2/s^2$).

The standard unit of *power* is the *watt*, symbolized W. One watt is the equivalent of one joule of energy expended per second of time (1 J/s). Power quantifies the rate at which energy gets produced, radiated, dissipated, or consumed. The faster we create or burn up energy, the greater the power.

TIP ***Various units smaller or larger than the watt can, and often do, express power. A* milliwatt *(mW) equals 0.001 W. A* microwatt *(μW) equals 0.000001 W or 10^{-6} W. A* nanowatt *(nW) equals 0.000000001 W or 10^{-9} W. A* kilowatt *(kW) equals 1000 W. A* megawatt *equals 1,000,000 W or 10^6 W. A* gigawatt *equals 1,000,000,000 W or 10^9 W.***

The standard unit of *electric charge quantity* is the *coulomb*, symbolized C. It's the amount of electric charge that exists in a "swarm" of about 6.24×10^{18} electrons. It's also the amount of electric charge contained in that same number of *protons*, *antiprotons*, or *positrons* (also called *antielectrons*). When you walk along a carpet with hard-soled shoes on a dry day, your body builds up a charge equivalent to a fraction of 1 C.

TIP ***The coulomb, like the mole, is a dimensionless unit. It represents a numerical quantity alone.***

The standard unit of *electric potential* or *potential difference*, also called *electromotive force* (EMF), is the *volt*, symbolized V. One volt is equivalent to one joule per coulomb (1 J/C). In real-world terms, the volt is a small unit of electric potential. A standard dry cell (of the sort you find in a flashlight) produces about 1.5 V. Most automotive batteries in the United States produce between 12 V and 14 V. Some high-powered radio and audio amplifiers operate with voltages in the hundreds or even thousands.

TIP ***You'll sometimes encounter expressions of voltage involving prefix multipliers. A* millivolt *(mV) equals 0.001 V. A* microvolt *(μV) equals 0.000001 V or 10^{-6} V. A* nanovolt *(nV) equals 10^{-9} V. A* kilovolt *(kV) equals 1000 V. A* megavolt *equals 1,000,000 V or 10^6 V.***

The standard unit of *electrical resistance* is the *ohm*, symbolized by the uppercase Greek letter omega (Ω) or sometimes written out in full as "ohm."

Resistance, which represents the amount of opposition that a substance (such as a wire or heating element) offers to electrical current, relates mathematically to the current and the voltage in any direct-current (DC) electrical circuit.

When a current of 1 A flows through an electrical component that has a resistance of 1 ohm, a voltage of 1 V appears across that component. If we double the current to 2 A and the resistance remains at 1 ohm, the voltage across the component doubles to 2V; if we reduce the current by a factor of 4 to 0.25 A, the voltage across the component drops by a factor of 4 to 0.25 V. In effect, an ohm represents a volt per ampere (1 V/A).

TIP ***The ohm is a small unit of resistance, so you'll often see resistances expressed in units of thousands or millions of ohms. A*** **kilohm** ***(k) equals 1000 ohms; a*** **megohm** ***(M) equals 1,000,000 (10^6) ohms.***

The standard unit of *frequency* is the *hertz*, symbolized Hz. Prior to about 1970, engineers called it the *cycle per second* or simply the *cycle*. If a wave has a frequency of 1 Hz, it goes through one complete cycle every second. If the frequency equals 2 Hz, the wave goes through two cycles every second, or one cycle every 1/2 second. If the frequency equals 10 Hz, the wave goes through 10 cycles every second, or one cycle every 1/10 second. The hertz can express sound-wave frequencies, such as the pitch of a musical tone. The hertz can also express or define the frequencies of wireless signals. You've probably heard the term in regards to computer microprocessor speed as well.

TIP ***The hertz is a small unit in the real world; 1 Hz represents an extremely low frequency. More often, you'll hear about frequencies in thousands, millions, billions (thousand-millions), or trillions (million-millions) of hertz. These units go by the names*** **kilohertz** ***(kHz),*** **megahertz** ***(MHz),*** **gigahertz** ***(GHz), and*** **terahertz** ***(THz) respectively.***

PROBLEM 10-7

A few years ago, people said that the clock speed of the microprocessor in the average computer doubles every year. Suppose that it's still true, and that it will remain true for at least a decade to come. If the average microprocessor clock speed today is 10 GHz, what will the average microprocessor clock speed be exactly three years from today?

SOLUTION

The speed will double three times, so it will grow to 2^3, or 8, times the value at which it started out. Therefore, three years from today, the average personal computer will have a microprocessor speed rated at 10 GHz × 8 = 80 GHz.

PROBLEM 10-8

In the preceding scenario, what will the average microprocessor clock speed turn out to be exactly 10 years from today?

SOLUTION

The speed will double 10 times, so it will become 2^{10}, or 1024, times as great as today's frequency. Computer microprocessors will (in this imaginary scenario) race along 10 GHz × 1024, or 10,240 GHz, by the time we finish our observations! We can also express this frequency as 10.240 THz, which we should round off to 10 THz because we know the original speed to only two significant figures. (Somehow I doubt that we'll ever see a microprocessor go that fast. What do you think?)

PROBLEM 10-9

Someone tells you that a radio transmitter has components that carry 3.0 kV of electric potential. How does that potential compare with the voltage produced by a flashlight type dry cell?

SOLUTION

The dry cell in a typical flashlight produces about 1.5 V. Because 1 kV = 1000 V, we know that 3.0 kV = 3000 V. That's roughly 3000 / 1.5, or 2000, times the voltage produced by the flashlight cell.

Unit Conversions

With all the different systems of units in use throughout the world, the business of conversion from one system to another has become the subject matter for numerous Web sites. Nevertheless, in order to get familiar with how units relate to each other, you should do a few manual calculations before going online and letting your computer take over. The problems below offer three simple examples. Table 10-1 can serve as a guide for converting base units.

TABLE 10-1 Conversions between common systems of units. Values with an asterisk (*) are approximate. All other values are exact.

To convert	Into	Multiply by	To go the other way, multiply by
meters (m)	nanometers (nm)	10^9	10^{-9}
meters (m)	micrometers (μ)	10^6	10^{-6}
meters (m)	millimeters (mm)	1000	0.001
meters (m)	centimeters (cm)	100	0.01
meters (m)	inches (in)	39.37*	0.02540*
meters (m)	feet (ft)	3.281*	0.3048*
meters (m)	yards (yd)	1.094*	0.9144*
meters (m)	kilometers (km)	0.001	1000
meters (m)	statute miles (mi)	6.214×10^{-4}*	1609*
meters (m)	nautical miles (nmi)	5.400×10^{-4}*	1852
kilograms (kg)	nanograms (ng)	10^{12}	10^{-12}
kilograms (kg)	micrograms (μg)	10^9	10^{-9}
kilograms (kg)	milligrams (mg)	10^6	10^{-6}
kilograms (kg)	grams (g)	1000	0.001
kilograms (kg)	ounces (oz)	35.28*	0.02834*
kilograms (kg)	pounds (lb)	2.205*	0.4535*
kilograms (kg)	English tons	0.001103*	907.0*
kilograms (kg)	metric tons	0.001	1000
seconds (s)	minutes (min)	0.01667*	60.00
seconds (s)	hours (h)	2.778×10^{-4}*	3600
seconds (s)	days (dy)	1.157×10^{-5}*	8.640×10^4
seconds (s)	years (yr)	3.169×10^{-8}*	3.156×10^7*
Kelvins (K)	degrees Celsius (°C)	Subtract 273*	Add 273*
Kelvins (K)	degrees Fahrenheit (°F)	Multiply by 1.80, then subtract 459*	Multiply by 0.556, then add 255*
Kelvins (K)	degrees Rankine (°R)	1.80	0.556*
amperes (A)	carriers per second	6.24×10^{18}*	1.60×10^{-19}*
amperes (A)	nanoamperes (nA)	10^9	10^{-9}
amperes (A)	microamperes (μA)	10^6	10^{-6}
amperes (A)	milliamperes (mA)	1000	0.001

TABLE 10-1 Conversions between common systems of units. Values with an asterisk (*) are approximate. All other values are exact. *(Continued)*

To convert	Into	Multiply by	To go the other way, multiply by
candela (cd)	microwatts per steradian (μW/sr)	1460*	6.83×10^{-4}
candela (cd)	milliwatts per steradian (mW/sr)	1.46*	0.683
candela (cd)	watts per steradian (W/sr)	0.00146*	683
moles (mol)	coulombs (C)	9.65×10^{4}*	1.04×10^{-5}*

Still Struggling

When you convert a quantity or phenomenon from one unit system to another, always make sure that you keep referring to the same quantity or phenomenon. For example, you can't convert meters squared to centimeters cubed, or candela to meters per second. You must keep in mind what you intend to express, and never attempt an impossible conversion task.

PROBLEM 10-10

Suppose you step on a scale and it tells you that you weigh 120 pounds. How many kilograms does that represent?

SOLUTION

Assume that you stand on the surface of the earth, so that you can define the mass-to-weight conversion function in a meaningful way. Use Table 10-1. Multiply 120 pounds by 0.4535 to get 54.42 kg. Because you know your weight to only three significant figures, you should round this result off to 54.4 kg.

PROBLEM 10-11

Imagine that you drive along a highway in Europe and you see that the posted speed limit is 90 kilometers per hour (km/hr). How many miles per hour (mi/hr) does this speed value represent?

SOLUTION

In this case, you only need to worry about miles versus kilometers; the "per hour" part doesn't change. You can simply convert kilometers to miles. First, remember that 1 km = 1000 m; then 90 km = 90,000 m = 9.0×10^4 m. The conversion of meters to statute miles (the miles that people use to express distances on land) requires that you multiply by 6.214×10^{-4}. Therefore, you multiply 9.0×10^4 by 6.214×10^{-4} to get 55.926 mi/hr. You must round this value off to 56 (two significant figures) because the posted speed limit quantity, 90, only allows you that much precision.

PROBLEM 10-12

How many feet per second does the above mentioned speed limit represent? Use the information in Table 10-1.

SOLUTION

You can convert kilometers per hour to kilometers per second first. This task requires division by 3600, the number of seconds in an hour. Thus, 90 km/hr = 90/3600 km/s = 0.025 km/sec. Next, convert kilometers to meters. Multiply by 1000 to obtain 25 m/s as the posted speed limit. Finally, convert meters to feet. Multiply 25 by 3.281 to get 82.025. You must round this value off to 82 ft/sec because the posted speed limit is expressed to only two significant figures.

QUIZ

Refer to the text in this chapter if necessary. A good score is eight correct. Answers are in the back of the book.

1. **If the resistance of a certain electrical device (such as a light bulb) stays constant while we decrease the current through it, what happens to the voltage across the device?**
   A. It stays the same.
   B. It increases.
   C. It decreases.
   D. We can't say because the premise of this question makes no sense.

2. **What's the difference between power and energy?**
   A. Energy depends on temperature; power doesn't.
   B. Power defines how fast energy gets produced or used up.
   C. Energy defines how fast power gets produced or used up.
   D. Power depends on temperature; energy doesn't.

3. **A nanogram (1 ng) equals precisely**
   A. 10^{-9} g.
   B. 10^{-6} g.
   C. 0.001 g.
   D. 1000 g.

4. **A microgram (1 μg) equals precisely**
   A. 10^{6} ng.
   B. 1000 ng.
   C. 0.001 ng.
   D. 10^{-6} ng.

5. **Suppose that we find a rock that weighs exactly 4.4 pounds. It has a certain mass. If we take that rock to Mars, where gravity is only about 0.37 times (37 percent) as strong as here on earth, we'll find that the *mass* of the rock has**
   A. not changed.
   B. increased by 37 percent.
   C. increased to 1 / 0.37, or 2.7, times its previous value.
   D. decreased to 37 percent of its previous value.

6. **Suppose that we remain on Mars and keep experimenting with our rock. We'll discover that, compared with the situation on earth, the *weight* of the rock has**
   A. not changed.
   B. increased by 37 percent.
   C. increased to 1 / 0.37, or 2.7, times its previous value.
   D. decreased to 37 percent of its previous value.

7. If someone tells you that a sample of water has a temperature of −15 K, you can conclude that

A. the water must exist in liquid form.
B. the water must exist as ice.
C. the water must exist as vapor.
D. the person doesn't know what he's talking about.

8. If we go exactly halfway around a complete circle 180°, what's the approximate angular measure in steradians (sr)?

A. 3.1416 sr
B. 4.9348 sr
C. 6.2832 sr
D. We can't say because the premise of this question makes no sense.

9. Approximately how many nautical miles (nmi) does a kilometer (1 km) represent?

A. 1.852 nmi
B. 1.609 nmi
C. 0.5400 nmi
D. 0.6214 nmi

10. Approximately how many micrograms (μg) does a pound (1 lb) represent on earth?

A. 4.535×10^{-9} μg
B. 2.205×10^{-8} μg
C. 4.535×10^{8} μg
D. 2.205×10^{9} μg

Final Exam

Do not refer to the text when taking this test. You may draw diagrams or use a calculator if necessary. A good score is at least 75 correct. The correct answer choices appear in the back of the book. It's best to have a friend check your score the first time, so you won't memorize the answers if you want to take the test again.

1. When we truncate the quantity 3.14159265 to two *significant figures,* we get

A. 3.142.
B. 3.141.
C. 3.14.
D. 3.1.
E. 3.0.

2. Consider the following sequences. In one of them, the right-hand quantity varies in inverse proportion to the left-hand quantity. Which one?

A. 1:2, 2:3, 3:4, 4:5, 5:6, 6:7
B. 1:2, 2:1, 1:3, 3:1, 1:4, 4:1
C. 64:10, 32:20, 16:40, 8:80, 4:160, 2:320
D. 1:5, 2:10, 3:15, 4:20, 5:25, 6:30
E. 8:7, 7:6, 6:5, 5:4, 4:3, 3:2

3. What mistake or formatting inconsistency, if any, have we committed in the following longhand multiplication problem?

			1	
		1	2	
		1	2	4
		×4	1	5
1	1			
		6	2	0
	+1	2	4	0
+4	9	6	0	0
5	1	4	6	0

A. We've done one of the single-digit multiplication steps wrong, and also carried a digit improperly.
B. We've done one of the single-digit addition steps wrong.
C. We've aligned one of the rows improperly, writing it down one decimal place too far to the left.
D. We've aligned two of the rows improperly, not going far enough to the left.
E. We haven't committed any mistake or formatting inconsistency.

4. In a pie graph, the sample proportions

A. always equal one another.
B. never add up to 100%.
C. sometimes, but not always, add up to 100%.
D. always add up to 100%.
E. never equal one another.

5. Figure Exam-1 shows a one-to-one correspondence between the

A. negative integers and the nonnegative integers.
B. counting numbers and the whole numbers.

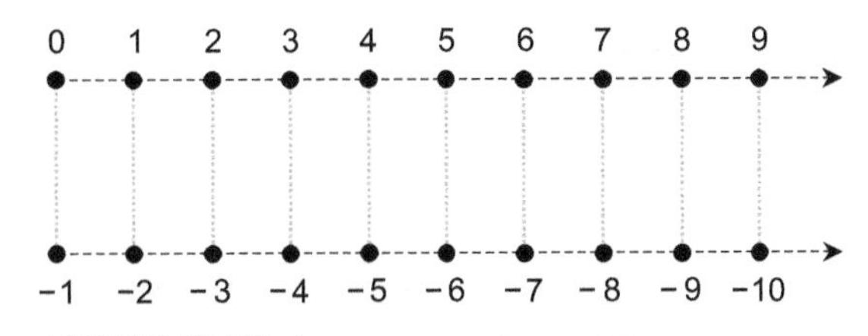

FIGURE EXAM-1 • Illustration for Final Exam Question 5.

C. rational numbers and the irrational numbers.
D. integers and the natural numbers.
E. integers and the whole numbers.

6. Energy can be expressed in terms of the

A. rate at which power dissipates per unit time.
B. ratio of current to resistance in an electrical circuit.
C. total accumulation of dissipated power over time.
D. ratio of resistance to voltage in an electrical circuit.
E. Any of the above

7. The Roman numeral MCCXIV represents the decimal quantity

A. 1214.
B. 1914.
C. 1203.
D. 1956.
E. 749.

8. When interest on a savings account gets paid at the end of every calendar year but never at any other time, the effective annual percentage rate (APR)

A. equals the net annual interest rate.
B. exceeds net annual interest rate.
C. is less than the net annual interest rate.
D. equals zero.
E. is actually negative.

9. Imagine that you come across the following endless decimal expression, where the number of 2s between the 5s keeps increasing by one as you move to the right:

$$0.25225222522225 \cdots$$

This quantity equals the fraction

A. 25 / 99
B. 225 / 999
C. 2225 / 9999
D. Any of the above
E. None of the above

10. A kilogram (1 kg) equals precisely

A. 100 milligrams (mg).
B. 1000 mg.
C. 10,000 mg.
D. 100,000 mg.
E. 1,000,000 mg.

11. In decimal numerals, we represent the Roman numeral XIX as

A. 9.
B. 11.
C. 19.
D. 21.
E. 99.

12. In a *horizontal* bar graph, the vertical heights of the bars depend on

A. the value of the independent variable.
B. the value of the dependent variable.
C. the value of the function at a specific point.
D. the average value of the function.
E. our preference (however tall we want to make them).

13. Which of the following numbers is prime?

A. 25
B. 21
C. 19
D. 15
E. 9

14. If we increase the radius of a perfectly round ball by a factor of exactly 10, we increase its *surface area* by a factor of

A. 10.
B. 100.
C. 100π, or about 314.16.
D. 1000.
E. 4000π, or about 12,566.

15. If we increase the radius of a perfectly round ball by a factor of exactly 10, we increase its *volume* by a factor of

A. 10.
B. 100.
C. 100π, or about 314.16.
D. 1000.
E. 4000π, or about 12,566.

16. Imagine that you have an unlimited supply of sugar cubes, all of which measure exactly one centimeter (a hundredth of a meter) along each edge. If you arrange

a bunch of them into a twenty-by-thirty-cube array, thereby getting a large, flat "sugar slab" measuring twenty centimeters wide by thirty centimeters deep by one centimeter tall, how many small sugar cubes do you have in total? Assume that you work in the decimal system.

A. 900
B. 600
C. 500
D. 300
E. 250

17. **If you neatly stack sugar cubes of the sort described above into a twenty-by-thirty-by forty-cube mass, thereby getting a "sugar block" measuring twenty centimeters wide by thirty centimeters deep by forty centimeters tall, how many sugar cubes do you have in total? Assume that you work in the decimal system.**

A. 24,000
B. 6000
C. 1200
D. 900
E. 800

18. **What mistake, if any, have we made in the following longhand addition problem?**

	9	8	7	6
	+6	0	4	1
1	5	8	1	7

A. We forgot to carry a digit from the ones place to the tens place.
B. We forgot to carry a digit from the tens place to the hundreds place.
C. We forgot to carry a digit from the hundreds place to the thousands place.
D. We committed both of the errors described in A and B, above.
E. We haven't made any mistake.

19. **Figure Exam-2 is an example of a**

A. horizontal bar graph.
B. paired-bar graph.
C. strictly nondecreasing graph.
D. vertical bar graph.
E. point-to-point graph.

20. **From the viewpoint of an engineer or experimental scientist, how (if at all) do the expressions 35.6 and 35.60 fundamentally differ?**

A. The first expression is less precise than the second one.
B. The first expression is more precise than the second one.
C. The first expression exceeds the second one by an order of magnitude.
D. The first expression is less than the second one by an order of magnitude.
E. An engineer or an experimental scientist would make no distinction between the two expressions.

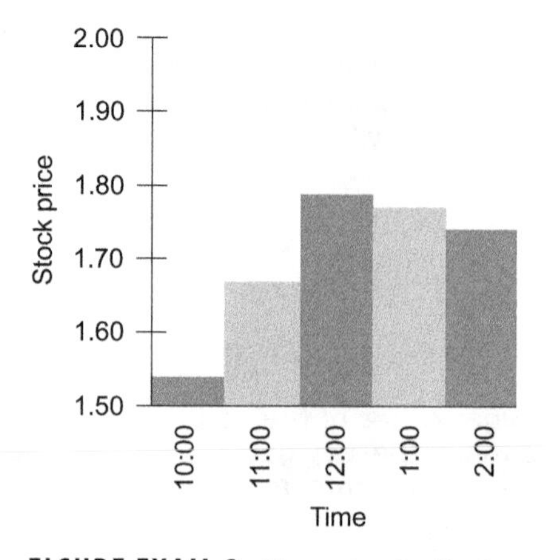

FIGURE EXAM-2 • Illustration for Final Exam Question 19.

21. We call a counting number composite when, but *only* when,

A. we can whole-number factor it into itself and 1, but in no other way.
B. it's odd, and it's also an even multiple of some other odd number.
C. it's even, and it's also an odd multiple of some other even number.
D. we can whole-number factor it into a product of two or more primes.
E. we can add it to some negative integer to get 0.

22. How would we write the quantity 378,034 in standard scientific notation, rounding off to three significant figures?

A. 378,000
B. 378×10^3
C. 3.78×10^5
D. 3.780×10^5
E. 37.8×10^4

23. A home-heating-system manufacturer tells you that a Model X gas furnace can provide 80,000 British thermal units per hour (Btu/h) of heating power for your home, while a Model Y wood-pellet stove can provide 32,000 Btu/h. What's the ratio of the Model Y stove's output to the Model X furnace's output?

A. 4:8
B. 3:5
C. 2:5
D. 5:8
E. 6:10

24. In the scenario of Question 23, what percentage of the gas furnace's output power does the wood-pellet stove provide?

A. 40%
B. 33%
C. 50%
D. 44%
E. We need more information to answer this question.

25. If we increase all three dimensions (the height, width, and depth) of a box by a factor of exactly 64, we increase its *surface area* by a factor of

A. 4.
B. 8.
C. 64.
D. 4096.
E. 262,144.

26. If we increase all three dimensions (the height, width, and depth) of a box by a factor of exactly 64, we increase its *volume* by a factor of

A. 4.
B. 8.
C. 64.
D. 4096.
E. 262,144.

27. If we cut an even counting number in half, we always get

A. another even counting number.
B. a nonwhole fractional quantity.
C. an irrational number.
D. an odd counting number.
E. an integer.

28. Suppose that we find a rock that weighs 10 pounds. It has a certain mass. If we take that rock to a planet where gravity is only half as strong as here on earth, we'll find that the *mass* of the rock has

A. grown larger by a factor of 4.
B. doubled.
C. not changed.
D. gone down to half of what it was.
E. gone down to 1/4 of what it was.

29. Which of the following expressions represents a quantity that differs from 785.64 by two orders of magnitude?

A. 7.8564×10^4
B. 78,564
C. 7.8564
D. 0.78564×10^1
E. All of the above

30. When we simplify the expression $0 + 3 \times 5 + 0 \times 6 + 7 + 0$, we get

A. 21.
B. 22.
C. 97.
D. 195.
E. 0.

31. When we simplify the expression $0 \times 3 + 5 \times 0 + 6 \times 7 \times 0$, we get

A. 21.
B. 22.
C. 97.
D. 195.
E. 0.

32. Where, if anywhere, have we made borrowing errors in the following longhand subtraction problem?

$$
\begin{array}{ccccccc}
0 & 8 & 15 & & & 4 & 12 \\
\not{1} & \not{7} & \not{5} & 7 & . & \not{5} & \not{2} \\
 & -2 & 8 & 5 & . & 4 & 7 \\
\hline
 & 6 & 7 & 2 & . & 0 & 5
\end{array}
$$

A. In the leftmost column only.
B. In the two leftmost columns.
C. In the third column from the left.
D. In the rightmost column only.
E. We haven't made any borrowing error.

33. Suppose that we find a rock that weighs 10 pounds. It has a certain mass. If we take that rock to a planet where gravity is only half as strong as here on earth, we'll find that the *weight* of the rock has

A. grown larger by a factor of 4.
B. doubled.
C. not changed.
D. gone down to half of what it was.
E. gone down to 1/4 of what it was.

34. Which of the following fractions constitutes a whole number?

A. 81/27
B. 91/13
C. 154/11
D. 171/19
E. All of the above

35. If someone tells you that a sample of water has a temperature of −10°C (degrees Celsius) on the surface of the earth, you can conclude that the

A. water is in liquid form.
B. water is frozen solid.
C. water is boiling.
D. water is entirely vapor.
E. person is misinformed.

36. Consider the decimal numeral 765.2751. If we move the decimal point three places to the right, we get a new quantity that's

A. three times the size of the original quantity.
B. one-third the size the original quantity
C. 1000 times the size of the original quantity.
D. 1 / 1000 the size of the original quantity.
E. the same size as the original quantity, but less precise.

37. If someone tells you that a sample of water has a temperature of +50°F (degrees Fahrenheit) on the surface of the earth, you can conclude that the

A. water is in liquid form.
B. water is frozen solid.
C. water is boiling.
D. water is entirely vapor.
E. person is misinformed.

38. In the binary numeral 101011, the digit on the extreme left represents

A. 256.
B. 128.
C. 64.
D. 32.
E. 16.

39. If you pay for a $4.66 item with a $10.00 bill, how much change should you get?

A. $2.33
B. $4.66
C. $5.34
D. $2.50
E. $14.66

40. When we want to multiply one fraction by another fraction, we multiply their numerators to get the new numerator, and we multiply their denominators to get the new denominator. Knowing this rule, we can show that the product of the rational numbers 0.111111··· and 0.222222··· is another rational number because it equals

A. $(1/9) + (2/9)$, which equals $3/9$ or $1/3$.
B. $(1/9) \times (2/9)$, which equals $2/81$.
C. $(1/9) \times (9/1)$, which equals $9/9$ or 1.
D. Any of the above
E. None of the above

41. When we want to add one fraction to another fraction that has the same denominator, we add their numerators to get the new numerator, and we use the denominator common to the two addends as the new denominator. Knowing this rule, we can show that the sum of the rational numbers 0.111111 ⋯ and 0.222222 ⋯ is another rational number because it equals

A. $(1/9) + (2/9)$, which equals $3/9$ or $1/3$.
B. $(1/9) \times (2/9)$, which equals $2/81$.
C. $(1/9) \times (9/1)$, which equals $9/9$ or 1.
D. Any of the above
E. None of the above

42. If we transpose the order of the quantities in a sum, the result

A. stays the same as it was before.
B. turns into the negative of what it was before.
C. turns into the reciprocal of what it was before.
D. becomes equal to 1.
E. becomes equal to 0.

43. If we transpose the order of the quantities in a product, the result

A. stays the same as it was before.
B. turns into the negative of what it was before.
C. turns into the reciprocal of what it was before.
D. becomes equal to 1.
E. becomes equal to 0.

44. Which of the following numbers is composite?

A. 5
B. 17
C. 37
D. 65
E. 73

45. When we simplify the expression 100 / 2 / 2 / 5, we get

A. 0.
B. 1.
C. 5.
D. 20.
E. an undefined quantity because this expression, as stated here, doesn't mean anything at all.

46. When we say that an operation between two quantities is commutative, we mean that

A. we have to write down the quantities in the correct order, if we expect to get the correct answer.
B. we'll get the correct answer regardless of the order in which we write down the quantities.

C. if we reverse the order in which we write down the quantities, we change the result into the negative of what it was before.
D. if we reverse the order in which we write down the quantities, we change the result into the reciprocal of what it was before.
E. the two quantities add up to give us 0 or multiply together to give us 1, or both.

47. A scientist would express or measure mechanical force in

A. joules.
B. kilograms.
C. seconds per meter.
D. newtons.
E. ohms.

48. Suppose that you buy a five-year certificate of deposit (CD) at your local bank for $5000.00, and it earns interest at a constant rate. At the end of the five-year term you cash it in and get $6150. By what percentage has the value of the CD increased over the entire five-year period?

A. 123%
B. 8.1%
C. 23%
D. 43%
E. 15%

49. Imagine that, instead of cashing in the CD at your bank as described in Question 48, you leave it in the bank for another five years at the same rate of interest. What will you get if you cash the CD in at the end of the second five-year term?

A. $7564.50
B. $7337.21
C. $7300.00
D. $7250.00
E. $7071.07

50. If we double the base length of a right triangle without changing the height, we increase the triangle's *interior area* by a factor of

A. 2.
B. 4.
C. 8.
D. 16.
E. 32.

51. If we double the height of a right triangle without changing the base length, we increase the triangle's *interior area* by a factor of

A. 2.
B. 4.
C. 8.
D. 16.
E. 32.

52. In the hexadecimal numeration system, what follows 99?

A. 100
B. 9A
C. A9
D. 101
E. Nothing because the digit 9 doesn't exist in the hexadecimal system.

53. If we increase the diameter of a circle by a factor of exactly 9, we increase its *circumference* by a factor of

A. 3.
B. 9.
C. 18.
D. 9π.
E. 81.

54. If we increase the diameter of a circle by a factor of exactly 9, we increase its *interior area* by a factor of

A. 3.
B. 9.
C. 18.
D. 9π.
E. 81.

55. The endless repeating decimal 0.27272727··· equals the fraction

A. 27/99
B. 2727/9999
C. 272,727/999,999
D. 27,272,727/99,999,999
E. All of the above

56. Figure Exam-3 is an example of

A. a pie graph.
B. a point-to-point graph.
C. linear interpolation.
D. a paired-bar graph.
E. linear extrapolation.

57. In a *vertical* bar graph, the vertical heights of the bars depend on

A. the value of the independent variable.
B. the value of the dependent variable.
C. the value of the independent variable minus the value of the dependent variable.
D. the average value of the function.
E. our preference (however tall we want to make them).

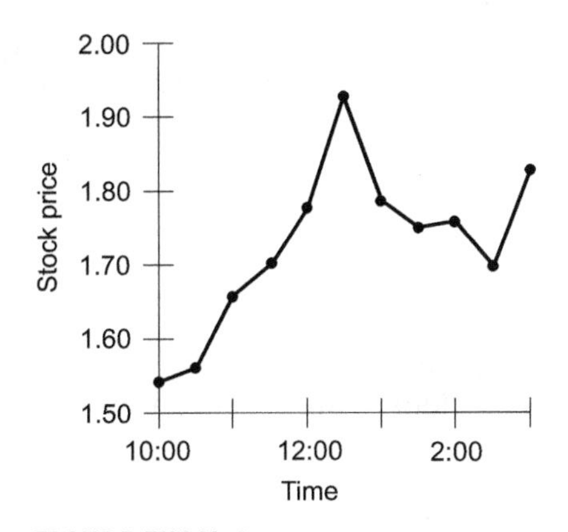

FIGURE EXAM-3 • Illustration for Final Exam Question 56.

58. How might a competent salesperson *correctly and properly* count back your change if you buy a $3.79 item with a $20.00 bill?

A. Hand you a penny and say "three-eighty," then hand you two dimes (10-cent pieces) and say "ninety, four," then hand you a $1.00 bill and say "five," then hand you a $5.00 bill and say "ten," and finally hand you a $10.00 bill and say "twenty dollars."

B. Hand you a penny and say "three-eighty," then hand you four nickels (5-cent pieces) and say "eighty-five, ninety, ninety-five, four" then hand you a $1.00 bill and say "five," and finally hand you three $5.00 bills and say "ten, fifteen, twenty dollars."

C. Hand you a penny and say "three-eighty," then hand you two dimes and say "ninety, four," then hand you a $1.00 bill and say "five," and finally hand you three $5.00 bills and say "ten, fifteen, twenty dollars."

D. Hand you a penny and say "three-eighty," then hand you two dimes and say "ninety, four," then hand you four quarters (25-cent pieces) and say "four twenty-five, fifty, seventy-five, five," then hand you a $5.00 bill and say "ten," and finally hand you a $10.00 bill and say "twenty dollars."

E. Any of the above

59. Consider the following sequences of ratios. In one, but only one, of these sequences, the right-hand quantity varies in direct proportion to the left-hand quantity. Which sequence is it?

A. 1:2, 2:2, 3:2, 4:2, 5:2, 6:2

B. 1:2, 2:1, 1:3, 3:1, 1:4, 4:1

C. 64:10, 32:20, 16:40, 8:80, 4:160, 2:320

D. 1:5, 2:10, 3:15, 4:20, 5:25, 6:30

E. 8:4, 7:5, 6:6, 5:7, 4:8, 3:9

60. What's the product $(7.47 \times 10^5) \times (4.238427 \times 10^4)$, taking significant figures into account, and taking precautions to avoid cumulative rounding errors?

A. 31.661×10^9
B. 3.17×10^9
C. 3.17×10^{10}
D. 3.166×10^{10}
E. $0.3166104969 \times 10^{10}$

61. What's the quotient $(7.47 \times 10^5)/(4.238427 \times 10^4)$, taking significant figures into account, and taking precautions to avoid cumulative rounding errors?

A. 17.624463
B. 17.62446
C. 17.624
D. 17.6
E. 18

62. What's the sum $(3.0007 \times 10^{66}) + (4.19 \times 10^5) + 66 + 2$, taking significant figures into account?

A. 3.0007×10^{66}
B. 3×10^{66}
C. 66
D. 68
E. 70

63. When you use a good personal computer's scientific calculator to determine sin 0.1000″ (the sine of one-tenth of an arc second with the angle expressed to four significant figures), you get

A. 4.848×10^{14}
B. 2.063×10^6
C. 4.848×10^{-7}
D. 1.000
E. 0.000

64. When you use a good personal computer's scientific calculator to determine sin 0.0100″ (the sine of one-hundredth of an arc second with the angle expressed to four significant figures), you get

A. 4.848×10^{15}
B. 2.063×10^7
C. 4.848×10^{-8}
D. 1.000
E. 0.000

65. If we move 1/4 of the way around the edge of a circle, what's the corresponding angular measure in radians (rad) at the circle's center, as we would measure it between our starting and finishing points on the circle's edge?

A. 2π rad
B. π rad
C. $\pi/2$ rad
D. $\pi/4$ rad
E. We can't say because the premise of this question makes no sense.

66. If we increase the height and width of a square by a factor of 5, we increase the square's *perimeter* by a factor of

A. 5.
B. 10.
C. 20.
D. 25.
E. 125.

67. If we increase the height and width of a square by a factor of 5, we increase the square's *interior area* by a factor of

A. 5.
B. 10.
C. 20.
D. 25.
E. 125.

68. What mistake, if any, have we committed in the following division problem?

$$
\begin{array}{rrrr}
 & & 4\,.\,0 & +r0.7 \\
\hline
16) & 6 & 4\,.\,7 & \\
 & 6 & 4\,. & \\
\hline
 & & 0\,.\,7 & \\
 & & 0\,.\,0 & \\
\hline
 & & 0\,.\,7 &
\end{array}
$$

A. We've made an intermediate multiplication error.
B. We've forgotten to carry a digit when necessary.
C. We've improperly borrowed digits when we shouldn't have.
D. We've made an internal division error.
E. We haven't committed any mistake.

69. You can minimize the total amount of interest that you have to pay on a long-term loan by

A. making the initial down payment as large as possible.
B. getting the highest APR that you can find.
C. paying off the loan as slowly as possible.
D. making the individual payments as small as possible.
E. All of the above

70. We can write the fraction 123,456 / 10,000 in decimal form as

A. 0.123456
B. 1.23456
C. 12.3456
D. 123.456
E. 1234.56

71. The following longhand multiplication matrix produces an incorrect answer. As we follow the steps in the conventional order in an attempt to find out why the result is wrong, what's the first error that we find?

	5	5	
	4	6	7
		×1	8
1			
3	7	0	6
+4	6	7	0
8	3	7	6

A. One of the single-digit multiplication steps is wrong.
B. One of the single-digit addition steps is wrong.
C. A carried digit is missing.
D. A borrowed digit is missing.
E. A remainder is missing.

72. Suppose that you own a garden that produces fabulous tomatoes. You sell your tomatoes to a wholesale distributor for $5.00 a dozen. The distributor sells the tomatoes to retail stores for $8.00 a dozen. The retail stores sell them to consumers for $14.00 a dozen. The federal government imposes a 12% value-added tax (VAT). How much VAT do you have to pay to the government when you sell a dozen tomatoes to a distributor?

A. Nothing
B. 36 cents
C. 48 cents
D. 60 cents
E. 72 cents

73. In the scenario of Question 72, how much VAT does the distributor pay to the government when it sells a dozen of your tomatoes to the retail store?

A. Nothing
B. 36 cents
C. 48 cents
D. 60 cents
E. 72 cents

74. In the scenario of Question 72, how much VAT does the retail store pay to the government when it sells a dozen of your tomatoes to a customer?

A. Nothing
B. 36 cents

C. 48 cents
D. 60 cents
E. 72 cents

75. Figure Exam-4 is an example of

A. a pie graph.
B. a point-to-point graph.
C. linear interpolation.
D. a paired-bar graph.
E. linear extrapolation.

76. Suppose that we have three identical pies and 15 people, all of whom tell us that they're hungry for pie. We want to give each person an equal amount, and we also want to make sure that we give each person the maximum possible amount given the quantity of pie that we have on hand. We should cut each of the three pies up into

A. thirds.
B. fourths.
C. fifths.
D. sixths.
E. sevenths.

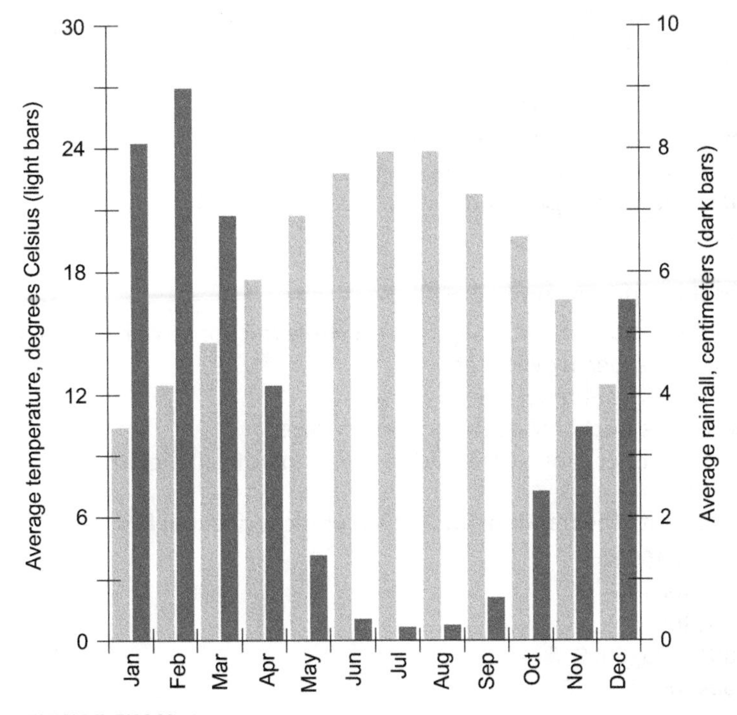

FIGURE EXAM-4 • Illustration for Final Exam Question 75.

77. Imagine that, after we've cut up the pies and served them to our 15 dessert guests in the scenario of Question 75, two of those guests decide that they don't want their pie after all. They give their slices of pie back to us. We're left with

A. 40% of a pie.
B. 60% of a pie.
C. 75% of a pie.
D. 80% of a pie.
E. a whole pie.

78. In the Roman numeration system, what symbol represents the *octal* quantity 10?

A. X
B. XII
C. XIII
D. XVI
E. VIII

79. In the Roman numeration system, what symbol represents the *hexadecimal* quantity 10?

A. X
B. XII
C. XIII
D. XVI
E. VIII

80. Under what circumstances would you want to *maximize* the APR for interest on a loan?

A. When it's a loan that the government makes to you
B. When it's a loan that you make to someone else
C. When it's a loan that a bank makes to you
D. When you want to pay it off as slowly as possible
E. Never

81. If we double a prime number, we'll get

A. a whole number in one case, and nonwhole fractions in all the other cases.
B. an even number in every case.
C. another prime number in every case.
D. a nonwhole fractional quantity in every case.
E. an odd number in one case, and even numbers in all the other cases.

82. We can an write the fraction 46 / 9999 in endless decimal form as

A. 0.004600460046 ···
B. 0.046046046 ···
C. 0.464646 ···
D. 0.000460004600046 ···
E. 0.46046004600046 ···

83. In Fig. Exam-5, curve X is

A. strictly nondecreasing.
B. infinite.
C. strictly nonincreasing.
D. value-added.
E. None of the above

84. In Fig. Exam-5, curve Y is

A. strictly nondecreasing.
B. infinite.
C. strictly nonincreasing.
D. value-added.
E. None of the above

85. In Fig. Exam-5, curve Z is

A. strictly nondecreasing.
B. infinite.
C. strictly nonincreasing.
D. value-added.
E. None of the above

86. Which of the following operations are commutative?

A. Addition and subtraction
B. Addition and division
C. Addition and multiplication
D. Subtraction and division
E. All of the above

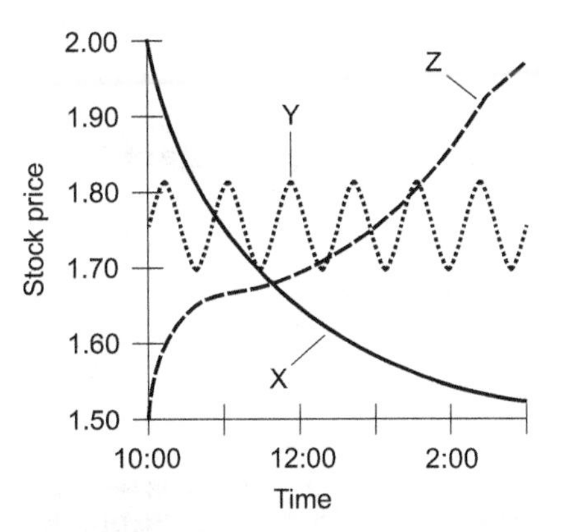

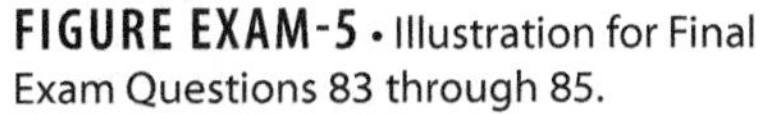

FIGURE EXAM-5 • Illustration for Final Exam Questions 83 through 85.

87. Suppose that our federal government enacts a flat income tax system with an exemption for the first $40,000 earned per year. If you make $100,000 in a given year and the tax rate is exactly 24%, what percentage of your income will you actually pay to the government in that year?

A. 12.0%
B. 14.4%
C. 16.0%
D. 18.8%
E. 20.2%

88. In the decimal numeral 13,865,427, the digit 3 represents

A. 300.
B. 3000.
C. 30,000.
D. 300,000.
E. 3,000,000.

89. A milligram (1 mg) equals precisely

A. 1000 grams (g).
B. 100 g.
C. 0.001 g.
D. 0.01 g.
E. 0.000001 g.

90. The fraction 7/5 is larger than the counting number 1 by an amount equal to one of the following percentages (rounded off to the nearest whole-number value). Which percentage?

A. 100%
B. 71%
C. 67%
D. 40%
E. 36%

91. The counting number 1 is a certain percentage of 7/5 (rounded off to the nearest whole-number value). Which percentage?

A. 100%
B. 71%
C. 67%
D. 40%
E. 36%

92. Suppose that you own a garden that produces wonderful carrots. You sell your carrots to a wholesale distributor for $1.00 a dozen. The distributor sells the carrots to retail stores for $1.80 a dozen. The retail stores sell them to consumers for $3.20 a dozen. The state government imposes a 5% retail sales tax. How much

of this tax do you have to pay to the government when you sell a dozen carrots to the distributor?

A. Nothing
B. 4 cents
C. 5 cents
D. 9 cents
E. 16 cents

93. Consider the decimal numeral 277,380.202. If we want to make this quantity two orders of magnitude smaller, which of the following actions can we take?

A. We can divide it by 100.
B. We can multiply it by 100.
C. We can move the decimal point two places to the left.
D. We can move the decimal point two places to the right.
E. We can do more than one of the above-described things.

94. Where, if anywhere, have we made borrowing errors in the following longhand subtraction problem?

	9	9	9		9	10
~~1~~	~~0~~	~~0~~	~~0~~	.	~~0~~	~~0~~
	−4	7	5	.	6	3
	5	2	4	.	3	7

A. In the leftmost column only.
B. In the leftmost and rightmost columns.
C. In all the columns except the leftmost one.
D. In all the columns except the leftmost and rightmost ones.
E. We haven't made any borrowing error.

95. If the resistance of a certain electrical device (such as a light bulb) stays constant while we decrease the voltage across it to 1/5 of what it was originally, what happens to the current that flows through it?

A. It stays the same.
B. It increases by a factor of 5.
C. It decreases by a factor of 5.
D. It increases by a factor of 25.
E. It decreases by a factor of 25.

96. Which of the following numbers would a mathematician call the smallest?

A. 7
B. −10
C. 13
D. −27
E. 0

97. When we *truncate* the number 245.89082 to *three significant figures*, we get

A. 245.
B. 246.
C. 245.890.
D. 245.891.
E. None of the above

98. When we *round off* the number 245.89082 to *three significant figures*, we get

A. 245.
B. 246.
C. 245.890.
D. 245.891.
E. None of the above

99. In the octal numeral 2745, the digit 4 represents the decimal quantity

A. 16.
B. 24.
C. 32.
D. 44.
E. 48.

100. Which of the following characteristics does a point-to-point graph *always* have?

A. The independent variable runs along the horizontal axis, increasing in value as we move toward the right.
B. The dependent variable runs along the vertical axis, increasing in value as we move upward.
C. It offers more precision than, or at worst the same precision as, a continuous-curve graph showing the same function.
D. It offers less precision than, or at best the same precision as, a continuous-curve graph showing the same function.
E. It reveals more detail than a horizontal bar graph showing the same function.

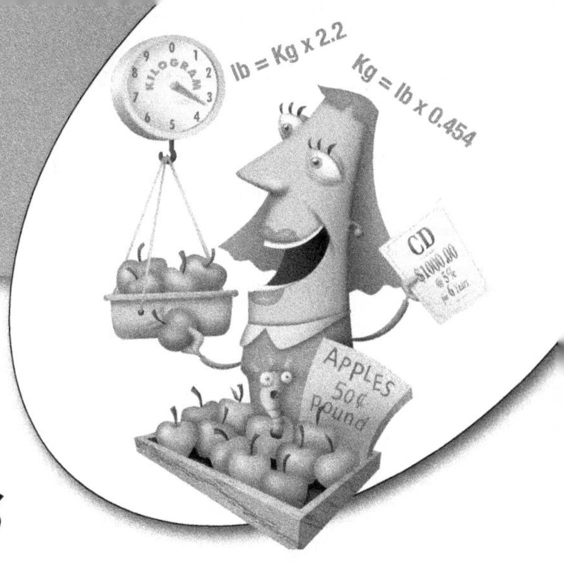

Answers to Quizzes and Final Exam

Chapter 1
1. D
2. B
3. A
4. A
5. C
6. B
7. A
8. B
9. C
10. D

Chapter 2
1. C
2. D
3. B
4. A
5. C
6. B
7. A
8. D
9. D
10. C

Chapter 3
1. D
2. C
3. B
4. A
5. C
6. B
7. C
8. D
9. D
10. D

Chapter 4
1. C
2. B
3. B
4. A
5. D
6. B
7. B
8. B
9. C
10. C

Chapter 5
1. B
2. C
3. B
4. A
5. B
6. A
7. D
8. D
9. C
10. A

Chapter 6
1. D
2. B
3. C
4. C
5. A
6. B
7. A
8. A
9. D
10. B

Chapter 7
1. B
2. D
3. A
4. D
5. B
6. D
7. C
8. A
9. C
10. B

Chapter 8
1. D
2. A
3. B
4. D
5. C
6. B
7. D
8. A
9. D
10. C

Chapter 9

1. C
2. B
3. C
4. A
5. A
6. C
7. D
8. B
9. A
10. C

Chapter 10

1. C
2. B
3. A
4. B
5. A
6. D
7. D
8. D
9. C
10. C

Final Exam

1. D
2. C
3. E
4. D
5. A
6. C
7. A
8. A
9. E
10. E
11. C
12. E
13. C
14. B
15. D
16. B
17. A
18. B
19. D
20. A

21. D
22. C
23. C
24. A
25. D
26. E
27. E
28. C
29. E
30. B

31. E
32. B
33. D
34. E
35. B
36. C
37. A
38. D
39. C
40. B

41. A
42. A
43. A
44. D
45. C
46. B
47. D
48. C
49. A
50. A

51. A
52. B
53. B
54. E
55. E
56. B
57. B
58. E
59. D
60. C

61. D
62. A
63. C
64. C
65. C
66. A
67. D
68. E
69. A
70. C

71. A
72. D
73. B
74. E
75. D
76. C
77. A
78. E
79. D
80. B

81. B
82. A
83. C
84. E
85. A
86. C
87. B
88. E
89. C
90. D

91. B
92. A
93. E
94. E
95. C
96. D
97. A
98. B
99. C
100. D

Suggested Additional Reading

Bluman, A., *Math Word Problems Demystified*, 2nd ed. New York, NY: McGraw-Hill, 2011.

Bluman, A., *Pre-Algebra Demystified*. New York, NY: McGraw-Hill, 2010.

Gibilisco, Stan, *Algebra Know-It-All*. New York, NY: McGraw-Hill, 2008.

Gibilisco, Stan, *Geometry Demystified*, 2nd ed. New York, NY: McGraw-Hill, 2011.

Gibilisco, Stan, *Mastering Technical Mathematics*, 3rd ed. New York, NY: McGraw-Hill, 2007.

Gibilisco, Stan, *Pre-Calculus Know-It-All*. New York, NY: McGraw-Hill, 2010.

Gibilisco, Stan, *Statistics Demystified*, 2nd ed. New York, NY: McGraw-Hill, 2001.

Gibilisco, Stan, *Technical Math Demystified*. New York, NY: McGraw-Hill, 2006.

Huettenmueller, Rhonda, *Algebra Demystified*, 2nd ed. New York, NY: McGraw-Hill, 2011.

Olive, Jenny, *Maths: A Student's Survival Guide*, 2nd ed. Cambridge, U.K.: Cambridge University Press, 2003.

Prindle, Anthony and Katie, *Math the Easy Way*, 2nd ed. Hauppauge, NY: Barron's Educational Series, 2009.

Ryan, Mark, *Everyday Math for Everyday Life*. New York, NY: Grand Central Publishing, 2002.

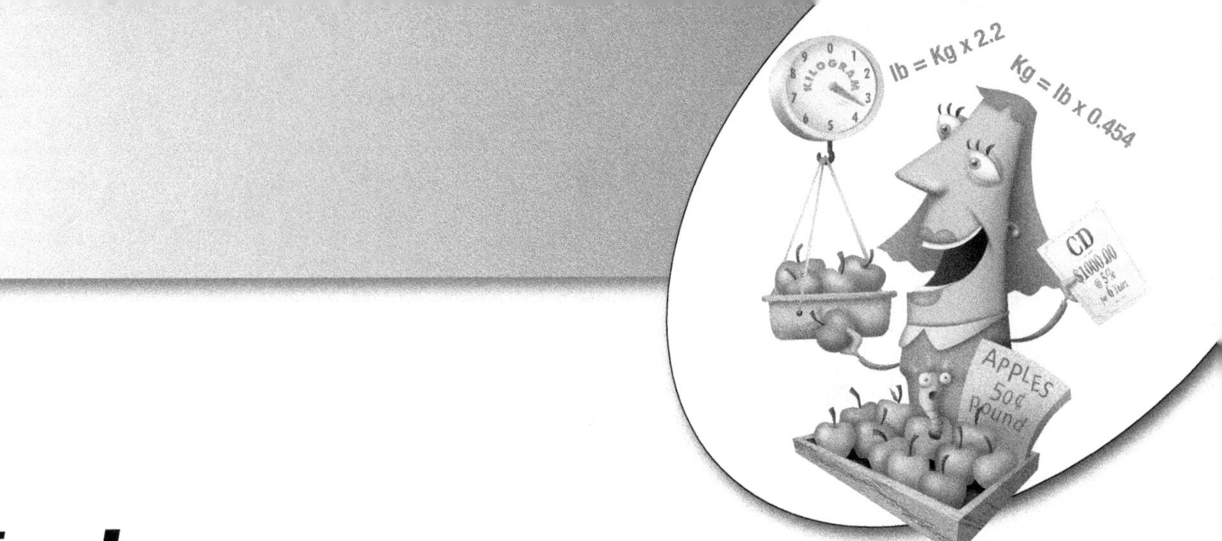

Index

C

G

H

I

JK

L

M

N

O

P

Q

R

S

T

U

V

W

XYZ

CPSIA information can be obtained
at www.ICGtesting.com
Printed in the USA
LVHW101512011020
667693LV00006B/213

9 780071 790130